Brakes and Friction Materials

Brakes and Friction Materials

The history and development of the technologies

Graham A Harper
CEng, FIMechE

with the support of Ferodo™

Ferodo Limited is part of T&N Friction **T&N**
Products Group
FERODO™ is a trademark of Ferodo Limited England

Published by
Mechanical Engineering Publications Limited
Bury St Edmunds and London, UK

First Published 1998

ISBN 1 86058 127 7

A CIP catalogue record for this book is available from the British Library.

Printed and bound in Great Britain by Bookcraft (Bath) Ltd.

Contents

Author's Note

Part One of this paper was originally presented at a joint meeting of The Institution of Mechanical Engineers and the Newcomen Society at the Institution in London on Wednesday 5th May 1971. The paper has now been brought up to date with the inclusion of Part Two. The complete paper will be reviewed at a joint meeting of The Institution of Mechanical Engineers, North Western Branch, the Newcomen Society, North Western Branch, and the Friends of the Museum at the Museum of Science and Industry in Manchester on Tuesday 2nd December 1997. The meeting is being held during the Ferodo Centenary Year 1897-1997.

G.A.Harper, CEng, FIMechE
October 1997

Preface to Part One

I would like to express my appreciation to the Institution of Mechanical Engineers and the Newcomen Society, for the honour of being allowed to present my paper, on behalf of the Institution, at the second Joint Meeting. Apart from my personal pleasure and satisfaction in carrying out this task, I feel that I am expressing a personal motive, since I was an enthusiastic supporter of the scheme for closer co-operation between the two organisations, when it was first discussed at the Institution. This same enthusiasm has now taken the form of a practical contribution, in gathering together in this paper, the results of my various inquiries made during the last seven years into the history of brakes and friction materials.

The subject of the paper is indicative of the fact that I am a member of both organisations, being fundamentally the history of the evolution of mechanical engineering over the centuries.

G.A.Harper, CEng, MIMechE (Member)
8th April 1971

Acknowledgements

The author wishes to thank the librarians of the Institution of Mechanical Engineers, the Newcomen Society, the Manchester Central Reference Library, the Patents Office, the Science Museum, the Assistant Secretary of the British Museum, the Public Relations Officers for Eastern, Western and London Midland Regions of British Rail, Westinghouse Brake and Signal Company, the "Autocar" and the Faculty of Technology in the University of Manchester for information and illustrations, and to colleagues for invaluable assistance in the preparation of photographs, slides and typing. Also the Directors of Messrs Ferodo Ltd for permission to publish data in connection with the development of friction materials.

The author would also like to thank the following for information used in the preparation of this paper :-

A.P.Lockheed, Leamington Spa, UK.

ADtranz, Derby, UK.

The Autocar, Haymarket Publishing, Teddington, UK.

The Buxton Advertiser, Derbyshire Times Newspaper Group, Chesterfield, UK.

Castrol Limited, Swindon, UK

Chapman and Hall, Andover, UK.

Crich Tramway Museum, Matlock, UK.

Dunlop Aviation Division, Coventry, UK.

Lucas Varity, Solihull, UK.

Raybestos, McHenry, Illinois, USA.

SAB Wabco BSI, Bromborough, UK.

Wabco Automotive UK Ltd, Leeds, UK.

Wabco Perrot Bremsen GMBH, Mannheim, Germany.

The following have given their assistance in the preparation and presentation of the material:-

The Library of the Institution of Mechanical Engineers, London, UK, for sourcing and loaning of books and technical papers.

Helen Harper for the typing.

Glenwood Studios, Chapel-en-le-Frith, Derbyshire, UK for photography.

Mechanical Engineering Publications Limited, Bury St. Edmunds, UK for the publishing.

T & N plc, Manchester, UK, for public relations in connection with the Ferodo Centenary Year.

The Museum of Science and Industry in Manchester, UK, for provision of presentation facilities.
The Directors of Ferodo Limited, Chapel-en-le-Frith, Derbyshire, UK, for permission to publish historical and current data in relation to Ferodo Limited, and to the many colleagues who have advised and assisted in so many ways.

Dedication

Herbert Frood 1864–1931

This book is dedicated to Herbert Frood.

Born 7th March 1864 at Balby, Yorkshire.
Died aged 67 in 1931 at Colwyn Bay.

Entrepreneur, inventor of synthetic friction material,
founder of Ferodo Ltd and the friction material
industry.

100 YEARS OF BRAKING HISTORY

1897 — 1997

FERODO
THE FIRST NAME IN BRAKES

FERODO LIMITED Chapel-en-le Frith, High Peak SK22 0JP, U.K.
Tel: +44 (0) 1298 811300 Fax: +44 (0) 1298 811319
FERODO is a trademark of Ferodo Limited

Related Titles

Title	Author/Editor	ISBN
An Engineering Archive Leather-bound Edition	Winterbone	1 86058 052 1
An Engineering Archive Hardback Edition	Winterbone	1 86058 053 X
History of Tribology 2E	D Dowson	1 86058 070 X
Handbook of Vehicle Design Analysis	J Fenton	0 85298 963 6
Analysis of Rolling Element Bearings	Wan Changsen	0 85298 745 5
Train Maintenance – Tomorrow and Beyond	IMechE Seminar	1 86058 095 5

For the full range of titles published by MEP contact:

Sales Department
Mechanical Engineering Publications Limited
Northgate Avenue
Bury St Edmunds
Suffolk
IP32 6BW
UK

Tel: +44 (0)1284 724 384
Fax: +44 (0)1284 718 692

Part One
Earliest Times to 1929

Introduction

The history of brakes is the history of man's fight to control power, and since the harnessing of power gave birth to engineering it is also a record of the material progress of mankind throughout the ages.

Brakes have been associated with movement from earliest times and it is a reasonably safe assumption that the first brake was the human hand, as utilised by the potter in ancient times to control the speed of his wheel. The potters wheel appeared about 3000 BC and consisted of a stone disc which was pivoted in a stone socket, the necessary rotary movement and braking being imparted by the potter's hands. By 2000 BC a second wheel had been introduced below the first, which was turned and braked by the feet, the first brakes could therefore be termed hand and foot brakes, terms which are associated with the modern motor car. A foot controlled potter's wheel is illustrated in *Fig. 1*.

Fig. 1 Potter controlling speed of kick wheel with foot, 1771. (Courtesy of Science Museum)

There is some evidence that rope brakes were in use about 100 AD , and since rope was introduced in 5000 BC it is possible that this type of brake was in use much earlier, maybe as an aid to controlling the rolling of felled trees down slopes. It is difficult to trace any form of vehicle braking by early civilisations, and it can only be assumed that this was carried out by control of the animals being used as motive power. This was very much the case with the Roman Chariot in which skilful manipulation of the horse was of paramount importance, although there is some evidence that a brake was used, consisting of a length of chain wrapped around the protruding wheel hub with one end anchored to the chariot body, a braking action being obtained by pulling the free end. The water wheel did not give rise to any braking equipment since the machinery could be controlled by adjusting the flow of water. The band brake which was an essential feature of the windmill, has been in use for 500 years and has numerous applications on modern equipment. It was the railway age that gave rise to the transport brake and it is interesting to note that the compressed air, vacuum, electro magnetic and hydraulic brake systems were all invented in this era. The motor vehicle age also gave rise to many new braking systems, and the ever increasing speed and complexity of modern transport make heavy demands upon the technical resources of both brake and friction material manufacturer, in order to ensure safe operating conditions for the vast transport system which we now accept as an everyday part of our lives.

Ref. (1)

Windmill Brakes

The exploitation of wind power by windmills gave rise to a very stringent set of operating conditions, since it was essential to bring the mill machinery to rest by some form of braking mechanism, against the driving power of the wind. The windmill originated in Persia in 7 AD and the first definitely accepted reference to a windmill in England is contained in the *Chronicle of Jocelyn de Brakeland* published in 1437 AD which describes the building of a windmill by Dean Herbert in his glebe land at Bury St. Edmunds in 1191 AD The first illustration of a windmill is that of a post mill which appears on a brass upon the tomb at Kings Lynn of Adam de Walsokne, Merchant and Mayor of Lynn, who died in 1349 AD Some indication of the type of brake utilised in windmills prior to 1500 AD was amply provided by Ramelli in his book, *Le Diverse et Artificiose Machine*, published in 1588. This work contains an illustration of a post mill utilised for grinding corn as reproduced in *Fig. 2*. The mill consists of a superstructure containing the grinding machinery and sails, the whole structure being pivoted on a central post to enable the sails to be set into the wind. The drive to the revolving grindstone is taken from the windshaft containing the sails, by means of the large wooden toothed wheel termed a brake wheel since around its periphery is positioned a clasp or band brake operated by a lever and a series of rope pulleys. It is not possible to see the type of band used or how it was applied, it can only be assumed that this was by some form of dead-weight held in the off position. It is possible to see in this illustration the development of the band brake from the rope brake.

The braking system which became typical on the British and Continental windmills is also illustrated in *Fig. 2*. The brake or gripe as it was termed consisted of a number of blocks of elm wood, which were connected together at their ends by means of iron plates and bolts, to form a flexible band which rested on the outer periphery of the wooden brake wheel . The two ends of the band were brought close together on one side of the wheel, the lower end being anchored and the upper one attached by means of a link to a heavy wooden beam which was pivoted at one end below the wheel. The weight of the beam was sufficient to apply the brake by tightening the band about the brake wheel, and was held in the off position by means of a peg which was fastened into its side and located in a swinging hook, being pivoted such that it tended to swing towards the peg. The brake was applied by means of a rope which was fastened to the free end of a lever and passed to a position where the miller could see the sails in order to stop them in the required position, the beam being quickly lifted by the rope such that the hook was swung clear of the peg and lowered before it could locate again. Since the beam had to be lifted to release the brake, the system was obviously designed on a fail safe basis by the millwrights.

The brake wheel was a very important item in the mill machinery, since apart from transmitting the power to the grinding stones, it was also subject to the full braking torque, which was considerable, as in the case of the North Mill, Wymondham in Norfolk, which is 9ft diameter and 8" wide. The brake wheels were originally of the type termed the compass arm, and consisted of two or three wooden arms mortised through the wooden wind shaft and secured with wedges, the rim being located on the end of each arm. This type of wheel eventually gave way to the clasp arm type, mentioned in the Dutch mill books from 1728 onwards, in which four arms were used forming a square at the centre, the wind shaft being passed through the square and keyed in position. This was a much stronger wheel since it was unnecessary to mortise the shaft and it was also much more convenient to true up. The arms were usually made of oak and the rim of elm, although eventually cast iron was used for both items, and in some cases the surface of the wheel was grooved to locate the band. Differing types of brake band were used comprising elm blocks fastened to an iron hoop, which was either in sections or a continuous length, iron hoops, elm blocks threaded onto a square section iron bar and backed with hoop iron, or as used in France, a flexible hardwood band faced with wooden blocks.

The windmill braking system remained unchanged to the end of the windmill era, although an interesting development was the introduction of a form of air brake, by a millwright from Suffolk called Catchpole about 1860. Essentially the brake consisted of a shutter arrangement fixed to the edge of the sails, which could be set normal to the direction of rotation of the sails giving a retarding effect.

Ref. (2), (3), (4), (5) & (6)

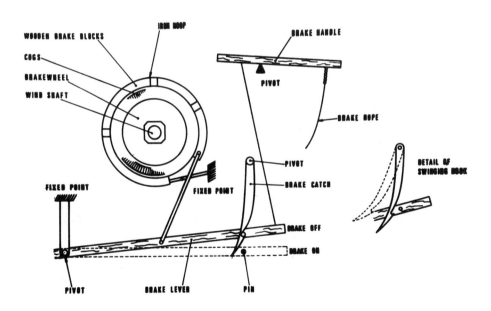

TYPICAL OPERATION OF BRAKE MECHANISM IN WINDMILL

Fig.2 Development of the windmill brake. A post mill prior to 1500. (Courtesy of the Science Museum) and the arrangement that became typical for western windmills

Mine Winding Engine Brakes

The mining industry has given birth to many engineering techniques and machines and this was very much the case in central Europe during the 16th century, as can be assessed from a description of a horse driven winding engine complete with a post brake, mentioned in a very comprehensive manual of mining technology published by Agricola, a German mining engineer, in 1556 *(Fig. 3)*. The winding gear consists of a horizontal winding drum, driven by means of wooden gears from a vertical horse driven shaft and due to the combined weight of loaded bucket and lifting chain, it was necessary to have some form of brake to safeguard the horse during ascent, and also to ease the load during the descent. This gave rise to the post brake which consisted of a wooden wheel fitted to the wooden winding drum shaft, onto the lower periphery of which was pulled a pivoted wooden brake block by means of a lever pivoted on a post, and connected to the block by a chain. The brake was applied by a brakesman situated near the top of the mine shaft who's job was to apply a load to the opposite end of the beam, thereby applying the brake block. It can be assumed that this type of brake developed from a much cruder technique of levering a wooden beam against the winding drum. These brakes were of considerable size with 15ft diameter wheels and brake blocks 10" square. The horse driven winding engines used in the British mines until the introduction of steam power in the 19th century did not in general use any braking system, but with the sinking of deeper mines a powerful post or cheek braking system became an essential feature of the steam winding engine. This consisted of two large vertical metal brake shoes lined with blocks of wood which were mounted such that the end of the grain clamped onto the winding drum, the brake shoes, which were pivoted below the drum, being applied by a steam operated bell crank and connecting rods, with manual operation available in the event of loss of steam. This type of brake which is also illustrated in *Fig. 3*, has remained almost unchanged apart from the introduction of synthetic friction linings in the early part of the 1900's, when lengths of rubber belting and a mixture of rubber belting and lead were used prior to the invention of impregnated cotton and asbestos friction materials.
Ref. (7), (8) & (9)

Fig. 3 Development of the mine winding brake. A post brake of 1555 (Courtesy of the Science Museum) and a modern colliery brake

Crane, Winch & Industrial Brake Systems

One of the earliest illustrations of a crane is found carved on the stone tomb of the Haterii, who were a family of building contractors living in Rome about 100 AD. The illustration shows a builders lifting pole operated by a tread mill, about the periphery of which is wound a rope, which apart from being used to assist in the winding, could also be used to control the lowering operation, being a simple form of rope brake, from which the true rope brake could well have developed as illustrated in *Fig. 4*.

Fig. 4 Relief from the tomb of the Haterii showing a rope to control a tread-mill powered builder's lifting pole, AD100. (Courtesy of the Science Museum)

An early example of winch braking is found on a number of the water wells in England during the 16th century, which were fitted with tread and donkey wheels for raising the water, braking being provided by a pivoted wooden beam which was pressed onto the underside of the tread wheel.

Due to the low speed and short work cycle associated with cranes and winches the band brake was ideal for this application, either lined with wooden blocks or used metal to metal. By 1912 block brakes, cone brakes and disc brakes were being utilised by crane builders.

Industrial braking or the control of machines by brakes, utilised various systems depending upon the power of the machines. Both band brakes and small post brakes were used, the latter eventually being utilised as safety brakes with solenoid operation for quickly stopping machines in the event of accidents.

Ref. (10), (11) & (12)

Railway Braking Systems

It is a fact that the first real development in transport brakes took place on the early railways, as will be seen from *Fig. 5*, which shows the tremendous rise in the number of railway brake patents from 1850. Railway systems were in use in the German mines during the 16th century, the wagon being guided by means of pegs which located in grooved planks, and the first reference to an English

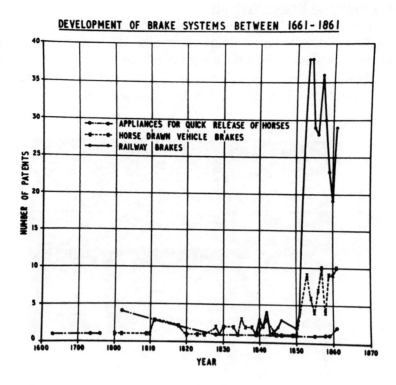

Fig. 5 The number of brake patents applied for between 1661-1861 indicating the impact of the railways

railway is made in a colliery account of 1597-8. With the quickly developing coal business, wagon ways with the wheels running in grooves were constructed to transport the coal from mine to the river or sea and by the end of the 18th century horse drawn wagons with flanged wheels running on the edge of iron rails was an accepted method of coal transport. The brake usually consisted of a wooden lever which was pressed downwards onto one of the wheels and many illustrations depict the lever in the shape of a tiller with the driver sitting on the end in order to apply the braking force. A more sophisticated system is mentioned by Desagulier in 1734 in his *Course of Experimental Philosophy*, as used by a Mr Ralph Allen on a railway to transport stone from his quarry to the River Avon at Bath, and consisted of a spragg arrangement fitted to the front wheels of the wagon and a "Jigg Pole" which was wound down onto the rear wheels by means of a windlass. A very interesting although simple braking system was used on the Peak Forest Tramway in Derbyshire which was constructed in 1797 and used for nearly 120 years to transport limestone from the quarries at Peak Forest to the canal at Buxworth, loads of 150 tons being moved in trains of 40 horse drawn wagons. Braking on the inclines was effected by thrusting an iron pin between the spoke wheels into a socket in the wagon frame, thereby causing the wheels to skid, the operation being carried out on as many wheels as considered necessary by a brakesman riding on the train, the pins which were fastened to the wagons with chains being released by means of a bar. The steepest incline on this particular tramway was fitted with a haulage system whereby a train of loaded trucks situated at the top of the incline, was connected by a wire rope which passed round a 14ft brake wheel to a train of empty trucks at the foot of the incline. The loaded trucks pulled the empty trucks up the incline, and control braking was applied by a band brake fitted with wooden blocks which clamped onto a smooth part of the brake wheel periphery. Another type of brake for use on horse drawn wagons was developed by a Mr Le Caan, a miner of Llanelly in Carmarthenshire in 1801, and consisted of a cast iron brake shoe which was pivoted above the wheel on the vertical centre line, the tip of the shoe almost reaching the rail, but kept clear of the wheel by means of a chain fastened to the pivoted horse shaft. In the event of the shaft being slightly lowered the shoe tip jammed between the wheel and the rail and the rolling motion of the wheel was changed to a sliding action. This brake is interesting since it is the first recorded instance of the use of cast iron blocks on railways.

The introduction of the steam locomotive during the 1820's did not bring about any development in systems of braking due to the poor mechanical efficiency of the locomotives. Some idea of the chronological development of railway brakes is given in *Table No 1*, and although many of the proposals did not get past the patent stage it is possible to detect the stages of development.

Chronological Development of Railway Brakes

Year	Inventor	Patent No.	Type of Brake	Where used (if known)
1801	L. Caan		Operated from the cart shaft by lowering of cast iron brake block on to rail.	
1831	G. Stephenson		Band brake on tender.	"John Bull" loco.
1832	G. Stephenson		Continuous system actuated from buffers.	Liverpool & Manchester Railway.
1833	R. Stephenson	6484	Steam operated on locomotive.	No record.
1836	H. Booth	6989	Closing exhaust pipe from cylinder on locomotive.	Liverpool & Manchester and in Belgium.
1839	J. Nasmyth	8299	Continuous system actuated by buffers.	
1840	B. M. Grover	8499	Electro-magnetic.	
1844	J. Nasmyth C. May	10358	Continuous system applied by vacuum.	Not used.
1848	J. Wilson		Continuous system, eccentric operated with cast iron blocks.	Colliery railway near Falkirk.
1848	R. Heath	12025	Continuous system, automatic, weight operated, held off by chain.	
1852	J. Newall		Continuous system, applied by springs.	East Lancashire Railway and France.
1853	E. Miles	444	Continuous system, applied by water pressure from loco boiler.	Shrewsbury & Hereford Railway.
1856	C. Fay	5002	Continuous system applied by screw.	Lancashire & Yorkshire Railway.
1857	A. Chambers W. B. Champion	385	Continuous system, applied by chain wound on axle driven drum.	North London Railway.
1860	J. McInnes	2594	Continuous system operated by compressed air.	Not used.
1862	J. Clark	2821	Continuous system operated by chain on axle driven drum in van.	
1864	C. Kendall	3083	Continuous system applied by compressed air.	London, Chatham & Dover Railway.
1869	E. D. Barker	587	Continuous system applied by hydraulic pressure.	Great Eastern Railway.
1869	G. Westinghouse	88929*	Continuous system applied by compressed air.	Panhandle Railroad.

The Two Main Developments

Year	Inventor	Patent No.	Type of Brake	Where used (if known)
1872	G. Westinghouse	(124404* (124405*	Continuous system, automatic, operated by compressed air.	
1878	J. A. Aspinall	4521	Continuous system, automatic, operated by vacuum.	Great Southern & Western Railway of Ireland.

* American Patent No.

Note: Table is only a selection of brakes and does not cover all the brakes developed or proposed.

Table 1. The development of railway brakes between 1801 and 1878 containing the fundamental principles of all future transport brakes

Generally speaking, up to 1850 braking was applied only on the locomotive or guards van at the rear of the train, the wooden brake blocks being applied by steam power on the locomotive and handscrew on the guards van. In 1831 George Stephenson fitted a hand operated brake to the tender of his *John Bull* locomotive and in 1832 developed a system for use on the Manchester to Liverpool Railway, in which brakes on a train of vehicles were applied when the buffers were compressed due to the initial braking of the locomotive, facilities being provided to adjust the brake load and also to make the system inoperative during shunting. Since the brakes were applied instantaneously throughout the length of the train, it was the first attempt to devise a continuous system and it is indicative of Stephenson's foresight that he realised the need for continuous braking at such an early stage in railway development. His son, Robert Stephenson, patented a steam operated locomotive brake in 1833, the brake blocks, or clogs as they were termed, being applied by a steam operated plunger working through suitable rigging. This brake did not materialise due to being applied only to one side of the locomotive giving rise to an off balance braking action. This was partially overcome by not having flanges on the wheels being braked, but another source of trouble was that the buffing equipment on the carriages or trucks was not suitable for the impact caused by the rapid application of the brake. A different type of locomotive steam brake was patented in 1836 by H.Booth, Secretary of the Liverpool and Manchester Railway which operated by closing the exhaust pipe from the cylinder thereby causing a back pressure and consequent retardation of the locomotive. A novel idea which was well ahead of its time was patented in 1840 by H.M.Grover and amounted to the first electro-magnetic brake, with the electro magnets suspended adjacent to the wheel surface which it was intended to grip when the magnets were energised from batteries in the brake van. A braking system patented in 1844 by J.Nasmyth of machine tool fame, and C.May proposed to utilise the power of vacuum, which again was an indication of things to come. An outstanding example of enterprise and invention is found in respect to a Mr J.Wilson who owned a colliery at Summerhouse near Falkirk in 1848. The colliery was connected by a railway to the Edinburgh and Glasgow Union Canal, and to which Wilson used to transport 24-cwt coal containers on trains made up of five trucks, braking being carried out on each wheel by cast iron blocks applied by eccentrics actuated by a bar running the length of the train, and controlled by a brakesman stood on a platform on the last truck. It is remarkable that the rail container system and a continuous braking system utilising cast iron blocks should have been in use at such an early stage of railway development.

Due to the greater train speeds, heavier loads and higher density of traffic, greater interest began to be taken by the various railway companies in the development of continuous systems, many of which were operated by the guard instead of the engine driver who was expected to give whistle signals in the event of an emergency. Due to the delay in operation, this system was the cause of many accidents, since the only brakes available to the driver were screw or steam brake. R.Heath patented in 1848 a weight operated system held in the off position by inclined planes, which could be either operated throughout the length of the train by rack and pinion gearing or from an inclined rail between the lines, thus providing a very early form of Automatic Train Control. This system was also used with the weights held in the off position by means of a chain, continuous throughout the length of the train, and which when released or broken applied the brakes, thereby forming the first automatic continuous braking system. The force released from wound up springs was used to actuate the brakes in a system patented in 1852 by J.Newall, Superintendent of the East Lancashire Railway, and the first hydraulic brake was patented in 1853 by E.Miles, water pressure being obtained from the locomotive boiler and transmitted to the brakes by means of a continuous pipe with flexible India rubber connections between the carriages. Mention is also made in this patent of utilising the hot water from the locomotive boiler for train heating purposes. In 1856, C.Fay of the Lancashire and Yorkshire Railway patented a braking system using worm and wheel gearing to apply the brake blocks, the guard turning the whole system throughout a train of carriages by means of a handwheel and flexible carriage joints. This system was used for over twenty years, the patent also covering the use of a "stand on the inside of the doors, of railway carriages for the facility of carrying umbrellas, sticks, canes and other similar articles". A system patented in 1857 by A.Chambers and W.H.Champion was power operated, the brake blocks being applied by means of chains wrapped around drums situated on each carriage which were belt driven from an axle, the drums being made to revolve when necessary by the guard through the action of a continuous belt striking gear. In 1858 the Board of Trade became interested in the number of railway accidents that were occurring due to lack of braking on trains, and the railway companies, in an effort to counteract the situation, added more brake vans to the trains. In order to reduce the weight and length of trains a composite brake carriage was introduced, with a small compartment for the guard and third class seating being built into one vehicle. The accidents still continued and more signal boxes were erected and the distant signals altered to give earlier warning to the driver. It had been a widely held belief amongst railway engineers, that the full braking effect from any system was only achieved when the wheels were locked and consequently skidding on the rails. This view had arisen as a result of the inadequate braking provided on trains before the development of the continuous system, which had consisted of braking on the locomotive and guards van only. Apart from being an erroneous assumption, it was also an expensive one as witnessed by the Manchester, Sheffield and Lincolnshire Railway, where due to the steep gradients, wheels were skidded for distances of four miles, resulting in the machining of tyres every two months at a cost of £2 and replacing every six months at a cost of £10. A realistic attempt to solve the problem was made by C.Fay in 1859 on the Goole branch of the Lancashire and Yorkshire Railway, using a modified version of his brake system. Instead of applying the brake blocks by means of gearing, the guard used a similar system to compress springs which could be released by removing a catch in the guards van. The test was carried out on three vehicles suitably ballasted to represent working conditions, which were slipped from the engine at 60mph, and the brakes applied, the vehicle coming to rest after skidding for 400 yards. The test was repeated after adjusting the springs, and the braking distance was 260 yards with no wheels skidding. As a result of these experiments the Lancashire and Yorkshire Railway prohibited wheel locking on its trains, but the practice continued generally for many years.

The first compressed air system was patented in 1860 by J.McInnes, the brake blocks being applied by means of a piston and cylinder with the air compressor and reservoir situated on the locomotive tender, the air supply being taken to each vehicle by means of pipes with flexible caoutchouc connections between the carriages. McInnes, who was a Glasgow engine driver was unfortunately unable to exploit his patent at this particular time, although it did appear in 1862 as the Steel and McInnes compressed air brake, and was used on the Caledonian Railway. A system by which the brake blocks were applied by a chain running the length of the train, which was tightened by wheel driven friction drums in the brake van was patented in 1862 by J.Clark, this system is important since it gave rise to two other systems, the first being developed by a Mr Naylor and used by the London Chatham and Dover Railway, utilised the chain to hold the brake blocks in the off position which gave fail safe working. The second was the Clark Webb system used by the London and North Western Railway for 25 years which used two brake blocks on each braked wheel instead of one, the chain being tightened as on the Clark system from the brake van. A quite successful compressed air brake was patented in 1864 by C.Kendall and used on the London Chatham and Dover Railway, with the compressed air at a pressure of 40lb/sq in. fed to the brake operating cylinders on each vehicle from an axle driven compressor situated in the guards van, by means of a continuous pipe which also acted as a reservoir. The air was admitted to the brake cylinders through valves actuated by chords throughout the train length, which were operated by either the guard or the engine driver. The connections between the carriages were flexible and designed to seal the air in the pipe in the event of a breakage thereby providing some element of safety. The patent also covered the use of an air operated whistle in the carriages for the passengers. A hydraulic brake patented in 1869 by E.D.Baker was used on the Great Eastern Railway for a number of years, hydraulic pressure being obtained from a hydraulic accumulator and fed to the brake cylinders by means of a valve which could be varied to suit adhesion conditions on the rails, the coupling hoses between the carriages being fitted with automatic valves to prevent loss of water. Although quite successful, the system was a superseded due to the inherent disadvantages of using water, especially in respect to freezing although brine was used to offset this tendency.

The 1870's were very decisive years in the development of railway braking systems, two systems being devised which were destined to become standard throughout the railway world, and two braking trials took place, which influenced the whole course of railway braking technology.

In 1870 an American engineer, Mr G.Westinghouse, became interested in railway brakes as a result of an accident on the Schenectady to Troy railway in 1866, his first design being of a buffer operated system, and a chain brake operated initially by a steam cylinder on the locomotive, and eventually by a steam cylinder on each carriage, supplied from the locomotive. After reading an article dealing with the use of pneumatic equipment in the Mount Cenis tunnel boring operations, he started to design a system utilising compressed air as the operating medium. This was patented in 1869, being a continuous non-automatic system with the brakes being operated by air cylinders on the carriages and tender, fed by a continuous pipe and flexible self sealing connections from a reservoir, steam pump and operating valve mounted on the locomotive. The system had been tried out on the Steubenville Accommodation on the Panhandle Railroad in 1868 and demonstrations were given on the Pennsylvania Railroad in September and November, 1869. The system was standardised and by 1st April 1874, the Westinghouse Air Brake Co., Pennsylvania, had fitted the system to 2,281 locomotives and 7,254 carriages. In July 1871, Westinghouse visited Europe in order to exploit the invention, but as far as England was concerned he had little success, finding it difficult to impress the railway managers. Whilst in England he met a Mr.Dredge of the *Engineering* journal, who showed him the draft of an editorial he was preparing in connection with railway air brakes, in which it was suggested that the brakes should operate in both parts of a train if accidentally separated, and that the failure of one carriage brake should not prevent the other brakes working normally. As a result of this meeting, Westinghouse patented an air brake system in 1872 which embodied these emergency features, and which he eventually developed into the standard continuous automatic brake used by American railway companies. The system consisted of a steam driven air compressor and main reservoir situated on the locomotive, which supplied compressed air through a drivers brake valve and a continuous train pipe to a valve, termed the triple valve, and auxiliary reservoir situated on each carriage, the auxiliary reservoir being connected to a brake operating cylinder, levers and brake blocks. The brakes were applied by reducing the standing pressure in the train pipe by means of the driver's brake valve, this operated the triple valve which shut off the compressed air from the train pipe, and allowed the compressed air from the auxiliary cylinder into the brake operating cylinder, thereby applying the brake blocks. To release the brakes the train pipe pressure was restored causing the triple valve to shut off the compressed air from the auxiliary reservoir, evacuate the brake cylinder, and admit compressed air from the train pipe into the auxiliary reservoir again for the next brake application. Provision was also made for the guard to apply the brakes as required by means of a valve, and in the event of any leakage in the system, or ruptures of the train pipe flexible connections between the carriages, the brakes were automatically applied. A diagrammatic layout of the Westinghouse system is shown in *Fig. 6*.

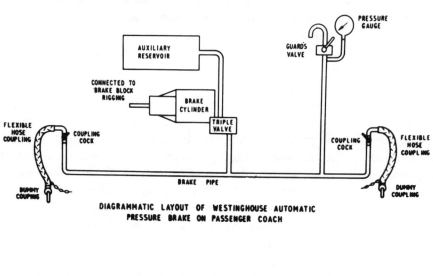

DIAGRAMMATIC LAYOUT OF WESTINGHOUSE AUTOMATIC
PRESSURE BRAKE ON PASSENGER COACH

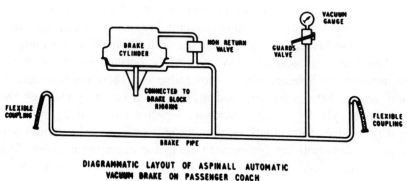

DIAGRAMMATIC LAYOUT OF ASPINALL AUTOMATIC
VACUUM BRAKE ON PASSENGER COACH

Fig. 6 The Westinghouse air pressure and the Aspinall vacuum systems which were to become standard throughout the railway world

Westinghouse also patented a locomotive steam brake in 1873, developed brake rigging, a communication system between the engine driver and the carriages worked by compressed air giving both audible and visual signals, automatic brake shoe adjusters and a vacuum operated brake system.

Although the matter of insufficient braking power had been taken up by the Board of Trade in 1858, the number of railway accidents after this date was sufficient to arouse public indignation, and this eventually led to the organisation of the Newark Brake Trials by the Royal Commission on Railway Accidents in 1875. The trials were carried out by a team of Royal Engineers on the Nottingham to Lincoln branch of the Midland Railway, and six companies were represented with eight braking systems utilising chain, compressed air, vacuum, hydraulic and hand operation, the best results being obtained from the Westinghouse automatic pressure system as used by the Midland Railway Company, on the basis of the amount of work done by each system per second.

The whole question of the fundamental aspect of railway braking assumed importance as a result of a visit made to England by Westinghouse in 1878. During the visit he attended a meeting at the Institution of Mechanical Engineers dealing with railway braking, and whilst taking part in the ensuing discussion offered to carry out railway brake trials under the supervision of a person to be nominated by the President of the Institution. This position was filled by Captain D.S.Galton, a man with a good deal of railway experience, and from May 27th - 29th 1878, a series of very important experiments were carried out which were destined to become the basis of all future railway brake design. The work was carried out in liaison with the London, Brighton and South Coast Railway who provided the locomotive *Grosvenor* and a special dynamometer van, designed by William Stroudley, the Locomotive Superintendent and Westinghouse, in which hydraulic cylinders and levers were utilised to obtain recordings of the force applying the brake blocks, the retarding force exerted by the blocks on the wheels, and the force required to drag the van with the brakes applied. A Westinghouse compressed air brake system was fitted to the van with cast iron brake blocks, the actual braking being carried out on one pair of wheels only, the speed of which was recorded on a self recording indicator designed by Westinghouse. In order to detect skidding a Stroudley speed indicator was attached to both the braked and unbraked axles, and the pressure from the hydraulic cylinders was recorded on four drums driven by a water clock.

Two basic types of experiment were carried out, the first being to drag the van at constant speed with the brakes applied and the second to uncouple the van from the locomotive whilst at speed, the brakes being applied automatically. The following three main conclusions were derived from the work :-

1. That the retarding effect of a locked wheel sliding on the rail was considerably less than when braked with a force that would just allow it to continue revolving.

2. That the coefficient of friction between the brake block and the wheel varied inversely to the speed of the wheel, the brake block requiring a higher pressure at higher speeds than lower speeds for the same retardation.

3. That the coefficient of friction of the brake block decreases a short time after the brake application has been made with a consequent loss of holding power.

In terms of railway brake technology the conclusions meant that it was wrong practice to skid the wheels on the rails to obtain maximum deceleration, it was necessary to proportion the brake block applying load in relation to the speed of the train to obtain a smooth retardation, and the phenomenon now termed brake fade was present after the brake block had been applied at speed.

Some further work was carried out by Galton on the North Eastern Railway in order to compare the brake application times of vacuum and compressed air operating systems, and although he had intended to investigate the retarding power of the various systems then in use, he did not pursue the work, considering it to be more the province of a Government Commission rather than a private individual. It is not difficult to imagine the delicate situation prevailing at the time with each of the many railway companies resolutely abiding by their own particular system, although from the point of view of the travelling public a good deal of tragedy would have been avoided if this work had been carried out. An interesting technical aspect of the Trials was the development by Westinghouse of an automatic brake pressure regulating valve, which ensured that the wheels did not skid, the idea was well ahead of its time and was not adopted by the railway companies.

Despite the obvious advantages of the compressed air brake it made little impression upon the British railway engineers, and further development took place in respect to vacuum operated brakes which was to become the standard one throughout the British railway system. This came about in 1878, when J.A.F.Aspinall of the Great Southern and Western Railway of Ireland patented an automatic continuous braking system operated by vacuum. A steam ejector on the locomotive creates a vacuum which is conveyed throughout the length of the train by a single pipe, to a brake operating cylinder unit and associated brake block levers on each carriage. One connection is made below the operating piston and one above the piston through a one way valve. In the running position a vacuum is created on both sides of the piston and the blocks are maintained in the off position. To apply the brakes atmospheric pressure is allowed into the train pipe by means of the driver's or guard's brake valve, which passes into the operating cylinder below the piston only, since the one way valve stops access above the piston. The pressure forces the piston upwards and applies the brake blocks through the various levers. The brake is automatic, since any ingress of air through the rupture of the flexible train pipe connections between the carriages, or any leakage into the system, will immediately apply the brakes. A diagrammatic layout of the Aspinall system is shown in *Fig. 6*.

The years immediately preceding 1890 were the closing years of a period of railway history often referred to as the *Battle of the Brakes*, for it was a period when railway men and many associated Board of Trade officials realised that despite the many braking systems in current use, it was becoming essential in terms of public safety to have a standard continuous brake in use throughout the railway system, like the Westinghouse system in America. It is a sad reflection on the people concerned that no legislation was brought to bear on this problem until the terrible Armagh Railway accident in 1889, in which eighty people lost their lives in runaway carriages fitted with non-automatic brakes. A great sense of urgency was created and by the end of the year the Government passed the Regulations of Railways Act, enabling the Board of Trade to enforce the use of continuous automatic brakes on the British railway system. The *Battle of the Brakes* was resolved and both vacuum and pressure automatic continuous brakes were fitted by the various railway companies during the next forty years, the ratio of vacuum to pressure being 2.7 to 1 by 1923, with the vacuum brake becoming standard after the railway amalgamation which took place during the same year. It is a matter of great interest why the British railway system should have developed the vacuum system when the merits of the Westinghouse air pressure system had been proved in England as early as 1875. Perhaps the reasons put forward by G.Prout in his *Life of Westinghouse* are pertinent ; he considered the reason was not the "invincible prejudice of the Britisher", but that the slower and more comfortable stop associated with the vacuum brake could be enjoyed in safety, since there were very few level crossings, the right of way was fenced from the ingress of people or animals, the lines were well signalled, there was a good sense of discipline amongst the railway staff and labour was cheap.

The braking on the locomotive was carried out mainly by using the boiler steam pressure which was fed by means of a brake valve to a brake cylinder unit connected to a system of rods and levers which forced cast iron brake blocks onto the driving wheels. The steam brake as fitted to the *King George V* is shown in *Fig. 7* where it is possible to see the cast iron blocks and lever system.

A very important aspect of railway braking is that of freight brakes, although this aspect was not developed to the extent of the passenger brake. Braking on freight trains was carried out for many years by the locomotive and the guards van, and when eventually a brake system was fitted to the wagons it was a very simple affair consisting of a hand operated lever and rigging, which applied one block to each pair of wheels. The braking procedure adopted was to apply the brakes at the top of the incline and release them at the bottom, the brake lever being held in the "off" or "on" position by pin and chain on a notched rack. Due to the lever being positioned on one side of the truck only, the system was the cause of many accidents to the railwaymen concerned, and in 1906 the Board of Trade Railway Employment Safety Appliances Committee subjected the brake to trials, and in a report published in 1907 condemned it as unsatisfactory. The matter was made a subject of law and in 1911 the Board of Trade enforced rules demanding that all new freight wagons must be fitted with hand levers on both sides of the wagon, and that the existing freight wagons were to be suitably modified within a specified time. Various designs of brake equipment were made to this specification in the early years of the twentieth century and it became standard throughout the British railway system. An interesting development which took place in America, as a result of freight brake trials on the Chicago, Burlington & Quincey Railroad at Burlington, Iowa, in 1886-87, was the development of a quick action triple valve. The trials proved that the current equipment was unsatisfactory but with the modified valves fitted, a train of fifty wagons was pulled up from 20mph in 200ft, the whole system being fully applied in 2½ seconds. As was the case with the braking systems fitted to passenger trains, there would seem to have been a more progressive approach in America toward freight brakes than that adopted by the British railway system.

Ref. (13), (14), (15), (16) & (17)

Fig. 7 The steam brake on the famous 'King George V' locomotive showing the cast iron brake blocks, levers, and pull rods. (Courtesy of E N Kneale)

Pedal Cycle Brakes

An interesting development in personal transport is that of the pedal cycle, which started in France when the Comte de Sivrac appeared at the gardens of the Palais Royal astride a small wooden horse fitted with two wheels, which he propelled and presumably braked with his feet. By 1860 bicycles, tricycles and quadricycles had appeared in a somewhat heavy form in Europe and America, but very few were fitted with brakes. A crank operated bicycle constructed by the Michaux firm in Paris in 1867 had a lever shoe brake fitted to it, which acted on the rear tyre, and was applied by means of a chord fastened to a revolving handlebar grip, as shown fitted with a wooden brake block in *Fig. 8*, the bicycle being termed a velocipede in France, and a boneshaker in England. In 1869 the first cycle show was held in Paris, and cycles were exhibited with front wheel brakes of the contracting band type, which is very important since this would seem to be the first application of the band brake to a vehicle.

During the next twenty years the cycle in the form of the ordinary or *Penny Farthing* bicycle became an accepted means of transport and considerable numbers were manufactured, especially in the Coventry area mainly due to the activities of the Coventry Machinists Co Ltd. Due to being able to apply a braking effect by back pedalling, other brakes were only a supplementary item and consisted of metal spoons or rollers applied as in previous models to the rear wheel tyre, by means of a cord attached to the handlebar grip. By 1880 the spoon and roller type brake was being fitted to the front wheel, operation being by handlebar levers or revolving handlebar grips and cam, and ground brakes were in use which acted on the ground from the rear of the cycle. An indication of future developments was the introduction of a caliper type brake by James Starley, a foreman at the

Coventry Machinist Co Ltd, who developed the tricycle, including a folding version amongst many other contributions to cycle development. In 1897 brakes which worked on the wheel rim were devised, the brake blocks consisting of leather or oiled compressed fibre, and freewheel brakes operated by a back pedalling action had been developed, this particular action being utilised by a form of roller clutch to apply either hub externally contracting, or internally expanding band brakes, and brakes acting on the wheel tyre.

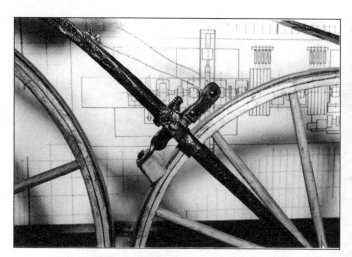 

Fig. 8 The close similarity between the bicycle brake on a Michaux boneshaker of 1867 (Courtesy of Ferodo Ltd) and a motor cycle of the early 1900s. (Courtesy of P Levesley)

By 1900 the cycle had developed into the conventional shape, using pneumatic tyres and speed gears, brakes mainly consisting of the spoon tyre applied by hand levers, but by 1915 stirrup-rim brakes, caliper rim and contracting band brakes operated by either handlebar mounted levers or Bowden type cables had been developed and were in use.
Ref. (18)

Motor Cycle Brakes

The motor cycle was a logical development of the pedal cycle, the first machine being a boneshaker fitted with a Perreaux single cylinder steam engine built in France in 1869. The brakes presumably would be the standard lever shoe brake acting on the rear wheel tyre, and operated from the handlebar grip as was also the first petrol engine motor cycle built by Daimler in 1885, which possessed a twist grip operated brake acting on an iron tyred wheel. After the passing of the Locomotives on Highway Act in 1896 which increased the maximum allowable road speed from 4mph to 12mph, motor cycle developments began to increase, and with the introduction of the light high speed petrol engine by Count Albert de Dion and Georges Bouton in 1895, numbers of motor tricycles fitted with the standard cycle brakes were constructed, which by 1900 were becoming very popular with the various English cycle manufacturers. Small motor cycle motors were being made in France and Belgium enabling the motorisation of the bicycle, and the Werner Brothers were developing, in France, belt driven motor cycles which by 1901 were being fitted with foot operated block brakes acting on the rear wheel belt rim. Brake operation by Bowden cable was used in 1903, and in 1905 the four cylinder F.N. was well ahead of its time, in having an internal expanding brake fitted to the rear wheel.

By 1908 braking systems consisted in general of front wheel braking by means of stirrup type brakes, and rear wheel braking with foot operated external contracting band brakes, or block brakes acting on the rim of belt driven models. *Fig. 8* shows a front wheel block braking system. These systems remained in general, unchanged for some twenty years, gradually giving way to front and rear internal expanding brakes using asbestos fabric friction materials, which due to being totally enclosed were not affected by wet road conditions. In common with many aspects of the motor cycle, a good deal of brake system development took place as a result of the many trials and race events, such as the Isle of Man T.T. Races where the machines were subjected to very exacting conditions.

Road Vehicle Brakes

As previously indicated, little attention seems to have been paid to brakes on horse drawn road vehicles, even the approved design of mail coach as detailed by the Commissioners of the Post Office Enquiry in their seventh report of 1837 did not possess brakes. The first reference is made in Thrupp's *History of Coaches* in 1877, which states that brake retarders were being fitted to the hind wheels to the exclusion of drag shoes. This is an interesting reference since drag braking in the form of a skid pan and roller scotch had been in use for many years. The skid pan was a cast iron pan which was placed under the front of the offside rear wheel prior to descending a hill, causing the wheel to slide on the cast iron surface thereby providing a braking action, the roller scotch was a round pin made from elm wood, which trailed along the road surface immediately behind one of the rear wheels when ascending a hill, preventing the vehicle from running backwards in the event of an emergency. When brakes were fitted, they simply consisted of brake blocks which were applied to the rim of the carriage wheels by means of pull rods and hand ratchet lever, the brake blocks consisting of elm wood, cast iron ,old rope, leather and camel hair belting or even an old boot. Some idea of the state of brake development in respect to horse drawn vehicles is given in *Fig. 5*.

The early development of mechanically propelled road vehicles was slow, despite the fact that Joseph Cugnot made a form of road traction engine as early as 1765, and by 1830 the steam coach had reached quite a high level of development, although braking was very simple consisting of a lever pressing on the rear wheels. Due to the state of the roads, the toll charges and competition from the railways, and especially the *Red Flag Act* of 1865, which enforced a speed limit of 4mph on open roads, and 2mph in built up areas, development in Britain came to a standstill. The Continent was not subject to speed restrictions and the first practical internal combustion engine road vehicle was developed in Germany by Karl Benz in 1885, consisting of a three wheel vehicle powered by a gas engine with a fast and loose pulley belt drive arrangement mounted on a countershaft, from which the road wheels were driven by chains. To retard the vehicle a small brake block was applied to the countershaft when the belt was running in the loose position. The introduction in 1884 of the lighter medium speed petrol engine by Daimler, and the light high speed petrol engine by Count Albert de Dion and Georges Bouton in 1895, enabled the commercial development of various forms of road vehicles, such that by 1900 road speeds in excess of 20mph were being attained. To control this speed pedal operated external contracting band brakes were fitted to drums on the sprockets of the rear wheel chain drive, or the simple hand operated spoon or block brake was used directly onto the solid tyre tread, in a manner similar to that of the horse drawn vehicles. The disadvantage with the band brake for road vehicle applications was its inability to function in reverse due to the loss of the self wrapping action, and to safeguard against runback whilst climbing hills, a sprag system was devised consisting of an iron spike which was hinged underneath the chassis, capable of being lowered onto the road surface at the correct angle, by means of an attached cord. A somewhat different design of brake appeared in France before 1898 designed by Messrs. Cloos and Schmalzer, and consisted of a brass lined ring which was expanded against the inside face of a metal collar. This is an example of a metal to metal brake which was apt to snatch when hot.

One of the most important developments in motoring history as far as Britain was concerned was the passing of the *Locomotives on Highways Act* in 1896, when the speed limit for motor vehicles of below 3 tons unladen weight was increased to 12mph. This had an immense effect both from a development and social aspect, with British engineers rapidly setting their own pace against their German and French counterparts, and the British public accepting the motor car as a cheap and reasonably reliable form of transport. The Daimler Motor Car Co. was formed in England in 1896 and produced the first car in March 1897, the first British designed car being that of F.W.Lanchester, who after a great deal of research had produced his first car in 1895, before the passing of the Act, incorporating a very original design of combined brake and cone clutch, in which the sliding part of the cone when released past the neutral position, could be brought to bear on a fixed cast iron ring providing a powerful braking action. Lanchester also patented a design of disc brake in 1902, consisting of a thin metal disc attached to the wheel hub of the vehicle, which was clamped during braking between two pieces of friction material fastened to pivoted lever arms operated by means of a pull rod. The design was well ahead of its time especially in respect to the friction material, which would be subjected to a very high operating duty, although Lanchester fitted oil immersed multi-disc brakes to 20hp and 25hp cars in 1906 which operated satisfactorily. Some indication of the type of braking systems in 1902 is given by the system fitted to the Daimler 12hp model, one of the most interesting features being the use of band brakes designed to be effective in both directions of rotation, by arranging for both ends of the band to be pulled simultaneously. A pedal operated two way external contracting band brake operated on a water cooled drum fitted to the rear of the gear box on a second motion shaft, and two similar band brakes operated on each rear wheel actuated by rods and a hand lever from the side of the car, with a balance beam fitted to ensure equal pull, all the brakes being linked to the clutch releasing mechanism to ensure that the clutch was released when the brakes were applied. The system on the Wolseley car was very similar, with the exception that the cars under 10hp possessed rear wheel brakes with blocks or shoes which were pressed against the inside of rims secured to the rear wheels, the shoes being lined with walrus hide.

The Automobile Club of Great Britain and Ireland which later became the Royal Automobile Club was formed in 1897, and competitions, reliability trials, hill climbs etc. were organised for the motoring public, from which came a good deal of technical development which in no small way affected brake design. One of the most outstanding developments was the introduction of the internal expanding brake by Mercedes in 1903, consisting of two brake shoes containing the friction material which were forced outwards against the inner walls of a drum fixed to the wheel, since the whole assembly was completely enclosed and was not therefore affected by dirt and water, it soon became adopted by vehicle designers. A typical internal expanding brake design is illustrated in *Fig. 9*. By 1905 the chain and belt drive on the lighter cars had given way to the universally jointed cardan shift drive from a gearbox to a differential rear axle, with a pedal operated band brake acting on the transmission shaft, and hand operated band brakes acting on the rear wheel hubs, this system gradually giving way to that of the internal expanding brake operated by pull rods.

Due to the inherent disadvantages of fitting brakes to the steerable front wheels, the two wheel brake system became standard practice until the early 1920's although a four wheel braking system was patented in 1904 by P.L.Renouf of Birmingham.

The development of commercial vehicle brake systems commenced in 1905 when the 12mph weight limit for vehicles was raised from 3 tons to 5 tons without trailer, and 6½ tons with trailer, thereby allowing heavy vehicle development to take place. The systems were of course based upon motor car experience, being made of larger proportions to handle the heavier loading.
Ref. (19), (20), (21) & (22)

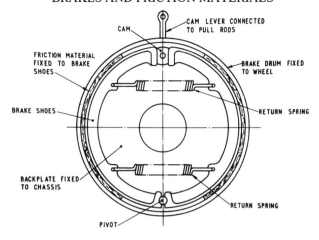

DIAGRAMMATIC LAYOUT OF INTERNAL EXPANDING BRAKE

Fig. 9 A diagrammatic layout of the internal expanding brake with the back face of the drum removed. Introduced by Mercedes in 1903 it became the standard vehicle brake

Brake Friction Materials

Although the vehicle designers produced braking systems which were capable of retarding the vehicle from higher road speeds, it soon became apparent due to the wearing and charring of the traditional friction materials used, that they were unable to withstand the higher duty. The materials used were elm wood, camel hair and cotton belting as used on the old horse drawn vehicles, but attempts were made to overcome the trouble by using metal braking in the form of cast iron or brass, and water cooled brakes. The problem was a serious one in respect to further brake system development, but was eventually solved by a Mr Herbert Frood, who in 1897 began to assess various types of friction material on a water driven dynamometer, situated in the hamlet of Combs, near Chapel-en-le-Frith, Derbyshire, as reconstructed in *Fig. 10*. This work eventually led to the manufacturing of a brake

Fig 10. In 1897 Herbert Frood made the first synthetic material in this hut, shown photographically reconstructed on a view of Frood's house at Combs near Chapel en le Frith, Derbyshire (Courtesy of Ferodo Ltd)

block built up from layers of textile material, impregnated with various ingredients, for which he was granted a patent in 1901, as illustrated in *Fig. 11*. This block was for use on horse drawn carriages, and by varying the ingredients could be used on either iron or rubber tyres, showing to good effect the capacity to vary the characteristics of manufactured friction material to suit the operating conditions. Despite the initial prejudice shown by the coachbuilders the brake block eventually became accepted and in order to satisfy demand Frood purchased two old cotton mills in 1902 at Chapel-en-le-Frith in Derbyshire, where he devoted himself to devising and manufacturing friction materials for automotive, railway and general engineering applications. In 1904 the London General Omnibus Company started to experiment with the use of motor buses, which were eventually introduced into service using Frood's cotton based friction material, satisfying the interesting requirement specified by Scotland Yard that the vehicle must be capable of pulling up in 14ft from the speed limit of 12mph. With the aid of test machines, a chemistry laboratory and a 6hp Benz motor car purchased in 1899, Frood was able to manufacture friction materials capable of withstanding the ever increasing demands of the vehicle manufacturers, which he covered by a patent granted in 1901 dealing with the impregnation of fabrics with high melting point varnish or enamel solution, the impregnated fabric being dried and left as a continuous roll or cut to the desired length and width to suit the various brakes. *Fig. 12* is an interesting period piece showing Mr Frood and his daughters attending a gymkhana at Chapel-en-le-Frith in 1906, being symbolic of the future growth of the friction material industry in respect to the automobile.

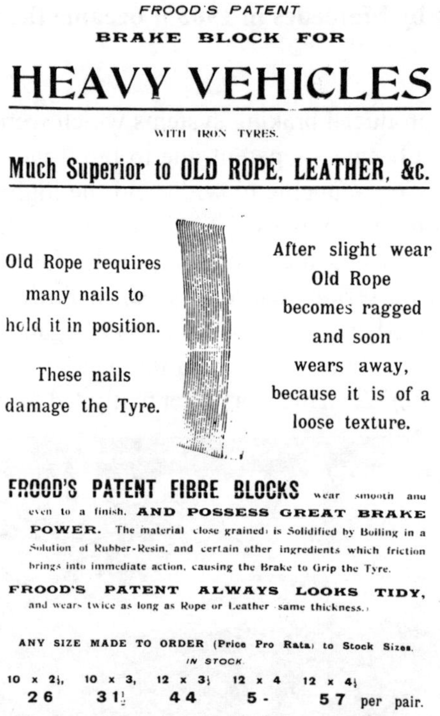

FROOD'S PATENT

BRAKE BLOCK FOR

HEAVY VEHICLES

WITH IRON TYRES.

Much Superior to OLD ROPE, LEATHER, &c.

Old Rope requires many nails to hold it in position.

These nails damage the Tyre.

After slight wear Old Rope becomes ragged and soon wears away, because it is of a loose texture.

FROOD'S PATENT FIBRE BLOCKS wear smooth and even to a finish. **AND POSSESS GREAT BRAKE POWER.** The material close grained is Solidified by Boiling in a Solution of Rubber-Resin, and certain other ingredients which friction brings into immediate action, causing the Brake to Grip the Tyre.

FROOD'S PATENT ALWAYS LOOKS TIDY, and wears twice as long as Rope or Leather same thickness.

ANY SIZE MADE TO ORDER (Price Pro Rata) to Stock Sizes.

IN STOCK.

10 x 2½,	10 x 3,	12 x 3½	12 x 4	12 x 4½	
2 6	3 1½	4 4	5 -	5 7	per pair.

Fig.11 Leaflet advertising Frood's patent brake block, 1904. (Courtesy of Ferodo Ltd)

One of the problems associated with the cotton based material was the charring and associated breakdown of the cotton fabric, when the brake linings were subjected to severe use such as continuous application on hill descents, a situation made even worse by the introduction of the totally enclosed internal expanding brake, which effectively sealed off the linings from the cooling air. This problem was solved by Frood, who in 1908 introduced an asbestos based friction material, a decision which was to influence the whole course of the friction material industry. The cotton fabric was replaced by asbestos fabric woven from long asbestos fibres spun into a continuous length on brass wire, the fabric being impregnated and dried in the same manner as the cotton based materials, this

Fig. 12 A technical period piece. Mr Herbert Frood and daughters symbolising in 1906 the age of the motor car at a local gymkana at Chapel en le Frith, Derbyshire, the home of friction materials. (Courtesy of Ferodo Ltd)

process forming the basis of all future woven material development. Although the asbestos based material proved very effective in overcoming the temperature problem, further complications arose due to the tendency of the new material to compress when the brakes were applied. This was eventually overcome by subjecting the material to a hot die pressing operation during manufacture, thereby increasing the density and removing the sponginess.

Future development took the form of strengthening the fabric by the introduction of various weaving techniques, using zinc wire in place of the brass or the removal of the wire altogether, developing new impregnants in an effort to increase the level of friction, and improving the penetration of the impregnant liquid into the fabric structure, by placing the fabric in steam heated tanks and creating vacuum conditions before introducing the impregnant, as illustrated in *Fig. 13*.

Although the main outlet for friction materials was the growing automotive market, Frood also became interested in railway braking as a result of the tragic Paris Metro accident which occurred on 10th August 1903, when eighty four people lost their lives in a train fire caused by a short circuit in a driving motor, due to the ingress of cast iron dust from the brake blocks. By 1907 a fabric type brake block had been developed for use by London Transport and the Paris Metro which not only overcame the fire hazard, but due to possessing superior frictional characteristics, enabled the train braking distance to be reduced, allowing an increase in the frequency of services.

Some idea of the growth of the friction material industry is gained from an examination of the output figures of one of the major Canadian asbestos mines, which show an increase from 8,772 tons in 1908, to 45,750 tons in 1928, a high proportion of this being used in the production of friction materials, which trebled in Canada between 1924 and 1929.

By the 1930's the staple products of the friction material industry consisted of brake and clutch linings for the automotive, motor cycle and engineering industries, moulded railway brake blocks and aluminium channel stairtreads. A certain amount of limitation was being placed upon the development of fabric material, due to the relative scarcity of the long fibre asbestos compared to the plentiful short fibre variety, the longer fibres being much easier to convert to warp in the spinning process. The situation in general was also much more complex, since due to the ever increasing vehicle loads and speeds, the demands made upon the friction material manufacturer were more stringent, whilst the scope offered to the manufacturer by the traditional combination of asbestos fabric and liquid impregnants only, was very limited, the use of powdered additives being prevented by the filtering action of the fabric structure. In an attempt to overcome this limitation a certain amount of development work was carried out by treating the asbestos yarn with emulsified powders prior to weaving, but this proved injurious to the textile machinery and eventually a completely new process was evolved, in which the short asbestos fibres were mixed with the required additives, and subjected to pressure in heated dies to form the required shape. This process completely revolutionised the industry enabling the manufacturer to use the short asbestos fibres in combination with whatever resins and additives were required, to produce a material of known friction and wear characteristics, the logical conclusion to a trend of events started by Frood in 1897, and which has become the basic process for the production of modern friction materials for drum and disc brakes. *Ref. (23) & (24)*

Fig. 13 Rolls of asbestos fabric being impregnated with chemicals as a stage in the manufacture of friction material during the 1930s. (Courtesy of Ferodo Ltd)

Conclusion

The history of brakes is an integral part of the history of mechanical engineering since whatever is made to go must also be made to stop. This has been a continual struggle over the centuries and as the foregoing pages testify, a good deal of mechanical ingenuity and skill has been exercised in solving the various problems, in an effort to keep pace with ever increasing speeds and loads. This situation still exists, and some idea of the high state of activity in this field is given by the fact that in 1970 there were 356 patents in the UK alone, whilst during the last decade the disc brake has become accepted on all forms of transport. Present day indications are that there will be no slackening of the pace in the future, in order to satisfy the requirements of the faster, heavier road vehicles, and the high speed trains planned for the worlds railways, which cannot even be conceived without braking systems to match the new operating conditions. Needless to say, there will be many changes in the traditional brake/friction material relationship as we know it, but the essential requirement will certainly not change – the power to stop.

Particular & General References

Introduction

(1)   "Brake Dynamics" by A.J.White, Motor Vehicle Research of New Hampshire, Lee, New Hampshire, USA.

Also of interest :

- "A short History of Technology from Earliest Times" by T.K.Derry and T.I. Williams, Oxford, 1960.

Windmill Brakes

(2)   "Notes on Old Windmills" by A.Titley, Transactions of the Newcomen Society, Vol 3.

(3)   "The Dutch Windmill" by F.Stokhuyzer, Merlin Press, London.

(4)   "The English Windmill" by R.Wailes, Routledge and Kegan Paul Ltd, London, 1954.

(5)   "Brake Wheels and Wallowers" by H.O.Clark and R.Wailes, Transactions of the Newcomen Society, Vol 16.

(6)   "Windmills - Their Rise and Decline" by R.Wailes, Engineering Heritage Vol 1, The Institution of Mechanical Engineers, London, 1963.

Also of interest :

- General Index to the Transactions of the Newcomen Society, Vols 1-30.

Mine Winding Engine Brakes

(7)   "The Winch from Well-Head to Goliath Crane" by F.R.Forbes Taylor, Engineering Heritage Vol 1, The Institution of Mechanical Engineers, London, 1963.

(8)   "Haulage and Winding Appliances used in Mines" by C.Volk, based on the works of Julius Von Hauer and translated by C.Salter, London, 1903.

(9)   "Winding Engines and Winding Appliances : Their Design and Economical Working" by E.McCulloch and T.G.Futurs, London, 1912.

Also of interest :

"Haulage and Winding" by Granville Poole, Ernest Benn Ltd, London, 1935.

Crane, Winch & Industrial Brakes Systems

(10)   "The Winch from Well-Head to Goliath Crane" See Ref (7).

(11)   "A 16th Century Treadwheel for Raising Water" by H.A.Sandford, Transactions of the Newcomen Society, Vol 4.

(12)   "Machine Design : Hoists, Derricks and Cranes" by H.D. Hess, London, 1912.

Railway Braking Systems

(13)   "Peak Forest Canal Tramway and Quarries" by G.Taylor, Great Central Railway Journal, Dec 1905.

(14)   "Continuous Railway Brakes" by M.Reynolds, London, 1882.

(15)   "Railway Carriages in the British Isles, 1830 - 1914" by Hamilton Ellis, G.Allen and Unwin.

(16)   "Effect of Railway Brakes" by D.Galton, three papers published in the Proceedings of the Institution of Mechanical Engineers, 1878 and 1879.

(17)   "Railway Mechanical Engineering" by A.R.Bell and Others, Gresham Publishing Co, London, 1923.

Also of interest :

- "Railway Brakes" by T.Rowatt, Transactions of the Newcomen Society, Vol 8. Excellent Bibliography.

- "North Western" by O.S.Nock, Ian Allan, London, 1968.

- "The Life of George Westinghouse" by H.G.Prout, American Society of Mechanical Engineers, 1921.

- "The Westinghouse Automatic Brake" Instruction Book, Westinghouse Brake and Signal Co Ltd, 1952.
- "The Galton Experiments" by J.G.Holmes, Railway Magazine, October 1961.
- "The Aspinall Era" by H.A.V.Bulleid, Ian Allen, London, 1967.
- "The Frozen Vacuum Brake" by J.G.Holmes, Railway Magazine, March 1961.
- "Britain's Railways Today" Edited by John St.John, Naldrett, London, 1954.
- "Red for Danger" by L.T.C.Rolt, David and Charles, Newton Abbot, 1971.
- Various Patents ref : Table 1.

Pedal Cycle Brakes
 (18)   "Cycles, History and Development" Part 1 by C.F.Caunter, HMSO, London, 1955.

Also of interest :
- "Free Wheels and Free Wheel Brakes" Proceedings of the Cycle Engineers Institute, Vol 1, 1899.

Motor Cycle Brakes
Also of interest :
- "Motor Cycles, History and Development" Part 1 by C.F.Caunter, HMSO, London, 1955.
- "Motor Cycles" Part 1. Historical Survey by C.F.Caunter, HMSO, London.
- "Motor Cycles" Institution of Automobile Engineers, London, 1918 - 1926.

Road Vehicle Brakes
(19)   "Railway Brakes" by T.Rowatt, Transactions of the Newcomen Society, Vol 8.
(20)   "Petroleum Motor Cars" by L.Lockert, Sampson Low, Marston & Co Ltd, 1898.
(21)   "Improvements in the Brake Mechanism of Power-propelled Road Vehicles" Provisional Specification by F.W.Lanchester, No 26407, dated 29th November 1902, The Patent Office.
(22)   "The History of Four Wheel Brakes" The Autocar, 19th November 1921.

Also of interest :
- "History and Development of Light Cars" by C.F.Caunter, HMSO, London, 1957.
- "Motor Cars" Part 2 by C.F.Caunter, HMSO, London, 1959.
- "Four Wheel Brake Developments" The Autocar, 17th August 1923.
- "Four Wheel Brakes" The Autocar, 16th and 23rd November 1923.
- "Four Wheel Braking History" The Autocar, 2nd and 9th January 1925
- "The Motor Car 1765-1914" by A.Bird, B.T.Batsford, London, 1960.
- "Automobile Brakes and Braking Systems" by P.T.Newcomb and R.T.Spurr, Motor Manuals No 8, Chapman and Hall Ltd, London, 1969. Contains chapter on the history of motor vehicle brakes.

Brake Friction Materials
(23)   "Bell Asbestos Mines Ltd 1878 - 1967" by G.W.Smith.
(24)   "Chrysotile Asbestos in Canada" by J.G.Ross, Canadian Dept of Mines No 707, 1931.

Also of interest :
- "The Ferodo Story" Ferodo Ltd, 1957.
- "Friction Materials, A Combination of Properties" Ferodo International technical News F.13.
- "Asbestos, from Rock to Fibre" by C.Z.Carroll-Porczynski, The Textile Institute, Manchester, 1956.
- "Asbestos, Its Industrial Applications" by D.V.Rosato, Reinhold Publishing Corporation, New York, Chapman and Hall Ltd, London, 1959.

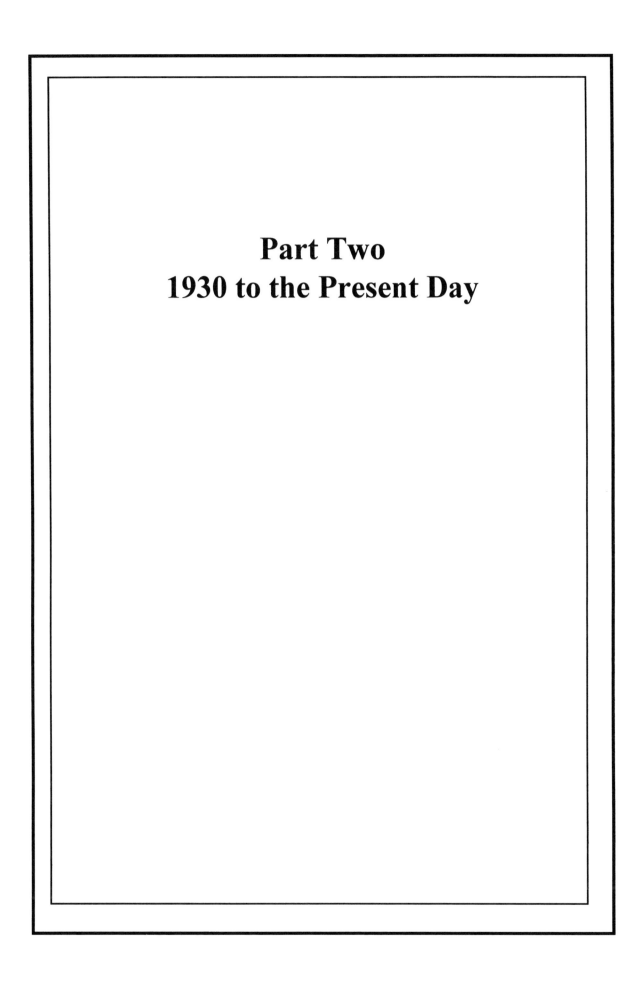

Part Two
1930 to the Present Day

Introduction

Part One, as published in 1971, dealt with the history of brakes and friction materials up to the early years of the present century. Since Part Two continues the history to the present day, it seems appropriate to blend the two sections together with a brief review of the main developments that have taken place, before proceeding in greater detail.

After a period of considerable expansion the various railway companies in the UK were amalgamated in 1923, and the Aspinall Vacuum braking system become standard on all passenger stock. In Europe and America the Westinghouse pneumatic braking system developed to suit the longer heavier passenger and freight trains. Both systems used cast iron brake blocks, and this situation remained unchanged until the introduction of pneumatically operated disc brakes and organic disc pads in the 1960's, with the increasing use of sophisticated actuating systems.

The brakes on road vehicles had been developed from the internal expanding drum brake introduced by Mercedes in 1903, operated mechanically by foot pedal through pull rods and levers or hydraulically. The brake shoes were lined with organic friction material developed on an organised chemical basis by Herbert Frood in 1897. Due to the increasing weight and speed of road vehicles a good deal of brake development and improvement had taken place, with many new designs being introduced as had occurred in the early days of the railways. Even as early as 1904 legislation was introduced demanding that motor vehicles should have two independent brakes. This was achieved by combining the band brake and the internal expanding drum brake, with the band clamping on the outside of the brake drum or the internal expanding drum brakes, being duplicated in the same brake drum either side by side or concentric.

As vehicle performance improved, and in order to avoid dangerous rear wheel skidding, a great deal of effort went into designing brake systems which could be fitted to the front wheels, compatible with the steering mechanism. This aspect was one of the most important developments in the history of brakes, since on four wheel braking systems the front brakes are subjected to far more work than the rear brakes. This is due to weight transfer from the rear axle to the front during the period of deceleration. With an even weight distribution this is in the ratio of 60:40, and has influenced the design of road vehicle braking systems throughout their development to the present day. From 1929, with the exception of one model, all British vehicles had front wheel braking systems, using the internal expanding drum brake, either mechanically or hydraulically operated, the rear brakes being also used as a parking brake operated from a separate handbrake lever.

In order to satisfy the ever increasing performance and speed of the private car, a number of different hydraulic internal expanding drum brake designs were developed in the immediate post war years. In the late 1950's the disc brake as originally patented by Lanchester in 1902, was gradually introduced using organic friction pads, eventually becoming a standard vehicle fitment, mainly on the front wheels.

Braking on commercial vehicles has been carried out with various designs of hydraulic and pneumatic internal expanding drum brakes. During the 1960's hydraulic disc brakes were fitted to the front wheels of the lighter models. This tendency has continued with pneumatically operated disc brakes being fitted on the front wheels of the heavier range of vehicles.

Motor cycle braking was carried out using manually operated internal expanding drum brakes, but with the great increase in performance, hydraulic disc brakes have become standard on the front and rear wheels. Cycling has become very popular in many forms, but braking is still achieved using the rim caliper brake with organic brake blocks.

During the 1930's aircraft braking, required during landing and taxying, was achieved using hydraulic internal expanding drum brakes. With the vast increase in the weight and speed of aircraft during hostilities a clutch type hydraulic disc brake was fitted, which has now developed into a specialised brake using carbon composite friction material.

With the increase in commercial and private transport there was a continuing demand for friction material in various forms, and this has led to the creation of a somewhat specialised form of industry both in the UK and world-wide.

In conclusion, the 1939-45 war effort produced a great deal of new technology which was exploited commercially for civilian use in the immediate post war years. This was very much the case with the transport industry, giving rise to a vast increase in the number, weight and speed of land based vehicles and aircraft. The associated demand for brakes and friction materials to match these new demands also produced many new designs and formulations. This occurred to such an extent that in the scope of this paper, it is only possible to review the main developments and follow the general overall trend up to present day requirements. The main items associated with these developments have been fully illustrated in relation to the text.

Internal Expanding Drum Brake

The chronological development of the internal expanding drum brake falls into three groups, leading-trailing, duo-servo and two-leading shoe.

(1) Leading-Trailing Shoe Brake   (Fig. 1)

This is the original design of brake introduced by Mercedes in 1903 and consists of a steel backplate fixed to the axle housing of the vehicle, on which are pivoted two brake shoes lined on the outer surface with friction material, which are expanded by a cam against the inner wall of a cast iron drum fixed to the vehicle wheel. The shoes are kept in contact with the cam by tension springs. Depending upon the rotation of the drum one of the shoes is pulled into the drum wall and is termed the leading shoe, whilst the other shoe is pushed away from the drum wall and is termed the trailing shoe, hence the term leading-trailing shoe brake. Since the leading shoe is pulled against the brake drum wall during a braking application, what is termed a self servo effect or wrapping action is created, greatly enhancing the performance of the brake.

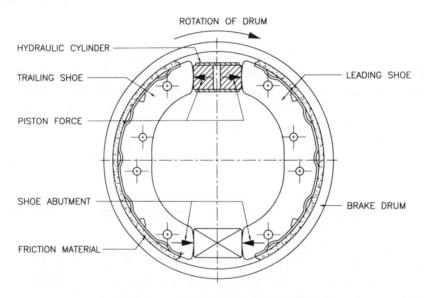

Fig. 1 Typical layout of a leading-trailing shoe brake

A very much improved version of this brake, as illustrated in *Fig.2*, was designed by the inventor Captain A.H.Girling in the early 1930's, the rotary cam was replaced by a very efficient cone type expander unit operated by a pull nod, which applied an expanding force to the brake shoes through rollers and tappets, thereby avoiding a sliding action. The expander unit was free to slide by a small amount on the backplate, enabling self centring of the shoes within the drum. The pivot for the brake shoes was designed in the form of a screw operated wedge, which could be adjusted from the backplate side to force the shoes apart to allow for the wear of the brake linings. The two tension springs, apart from keeping the shoes in contact with the expander unit, also held them against the steady posts fixed in the backplate, thus ensuring stability.

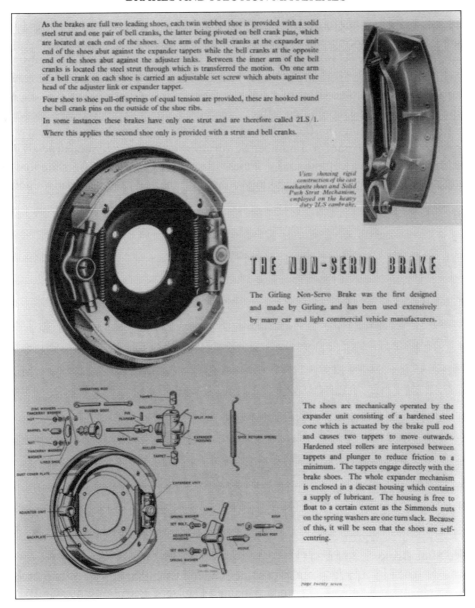

Fig. 2 Girling leading-trailing shoe brake. (Courtesy Lucas Varity)

This was a very successful design and after Pratt and Manley acquired the interests, was manufactured by the motorcycle firm New Hudson Ltd. in Birmingham, who became part of Joseph Lucas in 1943, and after combining with their Bendix brake company, became Girling Ltd. The design became the nucleus of a family of brakes produced by Girling after the war, including in 1945 a hydraulic version, with the mechanical expander unit replaced by a double acting hydraulic cylinder.

A hydraulic version was introduced by Lockheed for use on commercial vehicles, in the early 1930's, using a double acting cylinder. Towards the end of the 1930's another hydraulic version was designed for use on the rear wheels of commercial vehicles, with the cylinder replaced by what is termed a bisector. This consisted of a combined hydraulic cylinder and handbrake pull rod unit, which forced the brake shoes apart through an expander unit and tappets. It was fixed at right angles to the rear of the backplate enabling the hydraulic cylinder to be sited away from the hot brake drum. The handbrake pull rod was connected to the handbrake linkage, being operated independently from the hydraulic cylinder.

The leading-trailing shoe brake is a very versatile concept, since it can be used for both forward and reverse movement, and became the workhorse of the vehicle brake industry in various sizes for both front and rear brakes on motorcycles, cars, light and heavy commercial vehicles, with the rear brakes also being used as a parking brake, operated by a hand brake lever and cable.

Various modified forms were produced, such as the American Huck brake used by Bedford for their lorries, with the hydraulic operating cylinder functioning at an angle to the brake centre line, and the shoe pivot or abutment replaced by two links pivoted on a central anchor pin, each link being pivoted on the shoe ends, the combined effect creating a much increased self servo effect. As illustrated in *Fig. 3* by changing the flat sided operating cam used on the mechanical version to an "S" shape, or using a wedge expander a number of very powerful ruggedly constructed designs have been produced for use on heavy commercial vehicles. It is therefore very true to state that the original leading trailing design provides effective braking throughout the full range of road vehicles, from the private car to the heavy commercial vehicle.

Ref. (1) & (2)

Fig. 3 Girling leading-trailing "S" cam brake. (Courtesy Lucas Varity)

(2) Duo-Servo Shoe Brake   *(Fig. 4)*

This brake was designed by a French brake designer called Henri Perrot who at one time worked for the Argyll car company in Scotland. A company named Deutsche Perrot-Bremse GmBH was formed in 1926 to manufacture and sell the brake in Germany, and this company still operates at Mannheim.

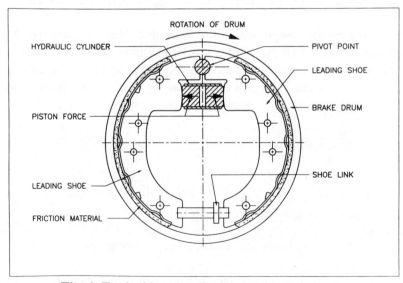

Fig. 4 Typical layout of a duo-servo shoe brake

Due to the considerable pedal effort required to brake the heavier cars, especially when brakes were fitted to all four wheels, Perrot designed an internal expanding drum brake, which produced more self servo effect than the equivalent size of leading-trailing shoe brake. The design consists of a backplate fixed to the axle housing, with two brake shoes lined with friction material connected at the bottom by a linkage. At the top of the backplate one of the brake shoes is pivoted and the other is actuated by a cam, the shoes being pulled together by tension springs. The relationship between the brake shoe pivot and drum rotation is such that, for forward rotation of the brake drum, the cam operated shoe is the leading shoe being pulled into the drum wall, thereby activating the opposite shoe into the leading mode through the linkage. This two-leading shoe effect produced a very powerful self servo or wrapping action, the resultant torque being constrained by the single pivot. The original design was very ineffective when used in reverse direction, since both shoes become trailing shoes, requiring a much increased pedal effort in order to provide braking. This problem was eventually solved by the brake designer V.Bendix, and consequently the Bendix Duo Servo brake became very popular on the large heavier cars used in America. Perrot Bremse designed an actuating system that enabled the brake to be fitted on the steerable front wheels, and in 1936 introduced a hydraulic version which was also effective in reverse, the linkage between the shoes being fitted with a screw type expander, to take up the wear of the brake linings. A modern version is illustrated in *Fig. 5*.

Fig. 5 Perrot duo-servo shoe brake. (Courtesy WABCO Perrot Bremsen GmbH)

Although the company does manufacture other types of brake, it has very effectively exploited the high self servo effect of the duo-servo shoe brake by accurate control and location of the brake shoes, and the use of brake linings having a stable level of friction over a wide range of operating temperatures, for use with lorries, buses and other heavy duty applications.
Ref. (3) & (4)

(3) Two-Leading Shoe Brake *(Fig. 6)*
Following the 1939-45 war great emphasis was placed by the car designer on body styling, the aim being to produce a more pleasing effect by streamlining the front and rear of the vehicle. This exercise created problems for the brake designer, since in order to carry out the new styling it was necessary to reduce the wheel diameter, and consequently limit the available diameter of the brake drum. Another associated problem was the introduction of independent front wheel suspension, encouraging the use of the far more compatible hydraulic brake system, rather than the mechanically operated system. A demand was therefore created for a hydraulic brake, which would provide more self servo effect than the leading-trailing shoe brake for the same drum diameter. It was to satisfy this demand that Lockheed introduced in 1947 the hydraulic two-leading shoe brake, which produced 50% more torque than the equivalent size of leading-trailing shoe brake. The fact that it was hydraulic is very significant, since Lockheed had been associated with hydraulic brakes since its inception, with

a connection going back to an engineer of Scottish origin called Malcolm Loughead, who in 1917 patented a practical hydraulic brake system. It is of interest that a vehicle had been manufactured as early as 1911 with mechanical brakes operated by hydraulic cylinders. The system known as the Weight Hydraulic System had a lever operated twin pressure unit or master cylinder, each cylinder operating one front and the opposite rear brake. The name Loughead became altered to Lockheed and the marketing rights were acquired by the Automotive Products Company in the early 1920's, and the Lockheed Hydraulic Brake Company was established at Leamington Spa to market brakes made in America. In 1924 Lockheed fitted the first hydraulic brakes used in the UK on a 12HP Bean motor car. Due to demand it was decided in 1929 to manufacture at Leamington Spa, and the now renamed A.P.Lockheed Company became the leading manufacturers in the UK of hydraulic brakes.

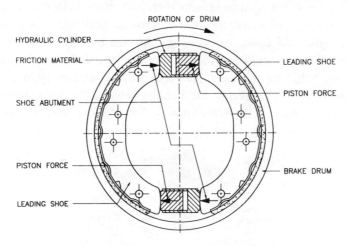

Fig. 6 Typical layout of a two-leading shoe brake

A mechanical version of the two-leading shoe brake had been used on racing cars before the 1939-45 war, and the design is basically a combination of the leading-trailing and duo-servo brake. The Lockheed version is illustrated in *Fig. 7* and consists of a backplate fixed to the axle housing, with a hydraulic cylinder and piston unit fastened in opposite directions at the top and the bottom. The two brake shoes lined with friction material are pivoted top and bottom at the rear of each cylinder unit, enabling the piston to push on the free end of each shoe, the shoes being pulled together by tension springs. A toothed cam and ratchet device was fitted between the end of the piston and the shoe tip, which could be adjusted through a hole in the backplate to take up wear of the friction material, and a small compression spring pressed on the shoe web to steady the shoe in relation to the backplate. As in the duo-servo brake the relationship between the brake shoe pivots and drum rotation is such that for forward rotation of the brake drum, both shoes are pulled into the drum wall and become leading shoes, this produces a powerful self servo effect, but since it is shared between the two shoe pivots is far more stable than the duo-servo brake. The brake is not suitable for reverse drum rotation, since both shoes become trailing shoes. This problem was eventually resolved by fitting the hydraulic two-leading shoe brake on the front wheels, and the hydraulic leading-trailing shoe brake on the rear wheels, which could also be used as a handbrake applied by a lever operated cable independently of the hydraulic actuation. The system was very well received by the post war car industry since it achieved balanced braking with light pedal effort, complementary to the new body styling,

Girling also introduced a hydraulic leading-trailing shoe brake in 1945 called the Hydrastatic. In order to keep the hydraulic brake fluid displacement to the minimum, springs were fitted in the brake assembly to remove any clearance between the surface of the friction material and the inside diameter of the brake drum. This facilitated the use of large diameter brake cylinder units enabling sufficient pressure to be applied with minimum pedal effort.

By 1950 nearly all the cars produced in the UK used brakes manufactured by Lockheed or Girling, with hydraulic two-leading shoe front brakes and leading-trailing shoe rear brakes becoming standard practice until the introduction of the disc brake.

Both Lockheed and Girling produced the two-leading shoe drum brake and a number of variants for use on commercial vehicles. They are of very rugged construction made in sizes of up to 15½" diameter by 6" wide (394mm x 152mm) for use with either hydraulic or pneumatic actuation.
Ref. (5) & (6)

Fig. 7 Lockheed two-leading shoe brake. (Courtesy A.P. Lockheed)

Relative Performance of Drum Brakes Described

In conclusion some idea of the relative performance of the internal expanding drum brakes described is given in *Fig. 8*, they can be categorised in descending order of torque as :-
(1) Duo-servo shoe brake.
(2) Two-leading shoe brake.
(3) Leading-trailing shoe brake.
Ref. (7)

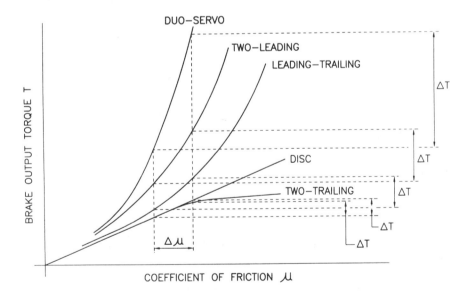

REPRODUCED BY KIND PERMISSION OF CHAPMAN & HALL.
AUTOMOBILE BRAKES & BRAKING SYSTEMS BY T.P.NEWCOMB & R.T.SPURR MOTOR MANUALS No.8 1969 PAGE 72.

Fig. 8 Variation of torque with μ for different brakes

Brake Drums & Brake Shoes

The early brake drums were made of steel having a relatively large working diameter in relation to a narrow rubbing path or wall, but with the introduction of the pneumatic tyre the diameter had to be reduced and the width of the rubbing path increased. Due to the increase in vehicle speed and the introduction of friction material containing metal wire, serious grooving and scoring occurred in the rubbing path, and drums manufactured from cast iron were developed. This was a very important development in inaugurating the traditional custom and practice relationship between synthetic friction material and cast iron, which is still part of modern braking technology. The excellent friction and wear properties of cast iron were even more enhanced by the development of Millenite and Chromidium special purpose high quality alloy cast irons in the 1930's.

Although the cast iron brake drum remained standard practice there have been variations to improve thermal efficiency, mainly for racing applications using an aluminium drum with a cast iron rubbing path. Cast iron drums have also been developed for high production use with an integrally cast thin steel back, providing a lighter drum with better thermal properties. Unfortunately, due to the natural shape of the brake drum it has never been possible to remove drum distortion in the form of bellmouthing, with the open unsupported end of the drum increasing in diameter during use. This produces uneven or taper wear of the brake linings which can cause serious braking and noise problems. Brake drums are produced in many sizes from the 7" diameter (178mm) drum used for small cars to the 15½" diameter (394mm) drum used for heavy commercial vehicles.

The shape of the brake shoe has basically remained unchanged, consisting of a shelf for the friction material with an underlying web for strengthening and rigidity purposes. The shelf contains a number of holes to enable the attachment of the brake lining by copper, brass or aluminium rivets, or is left undrilled when the brake lining is attached by bonding with high temperature adhesive.

Disc Brakes

(1) Automotive

The conception of the disc brake and internal expanding drum brake occurred at about the same time with Lanchester introducing the disc brake in 1902 and Mercedes the internal expanding drum brake in 1903. Although the internal expanding drum brake and its various forms became the accepted brake for road vehicles there was still further development of the disc brake. This development took two forms :-

1. Spot Type Disc Brake

A central disc with friction pads situated either side of the disc covering from 30° to 50° of the total 360° disc surface.

2. Clutch Type Disc Brake

A central disc with one annular disc of friction material situated either side of the disc covering the full 360° disc surface.

These developments occurred chronologically as follows :-

1906   Lanchester used multi-disc oil immersed clutch type disc brakes on his 20 and 25HP models operating on the rear of the worm drive.

1914   A.C.Cyclecar used clutch type disc brake with 10" diameter (254mm) discs operating on the end of the drive.

1923   Harper three wheel car used clutch type disc brake on the wheels.

1935   Dunlop Rubber Company designed clutch type disc brake for use on aircraft.

1937   Captain George Eyston fitted the first Lockheed hydraulic clutch type disc brake to his racing car Thunderbolt with Ferodo brake linings.

1937   Lockheed clutch type disc brake fitted to Airspeed Oxford aircraft.

1937   Hawley spot type disc brake fitted to Crosley car.

1939-46   A great deal of development was carried out with clutch type disc brakes for aircraft by Dunlop and Lockheed. Girling clutch type disc brakes were fitted to military vehicles.

1947   Dunlop introduce copper disc brake for aircraft.

1949   Spot type disc brakes fitted to wheels of Crosley car.

1951   Chrysler Corporation fitted clutch type disc brakes to the wheels of its larger models.

Although the internal expanding drum brake reigned supreme during the immediate post war years, a good deal of thought and consideration was given to the spot type disc brakes by Dunlop, Girling and Lockheed with a view to providing something suitable for road vehicles. As discussed previously, the streamlining of the vehicle bodywork had created a number of problems for the brake designer. Some of these problems had been overcome by the introduction of the hydraulic two-leading shoe brake, but there still remained the problem of brake overheating due to restriction of the cooling air by the bodywork. This caused brake drum distortion, overheating of the hydraulic brake fluid and loss of friction in the brake linings, termed brake fade. It was obvious that these problems would intensify with the improved performance and speed of vehicles, and that in the foreseeable future it would be necessary to carry out a fundamental change to the car braking system.

The use of clutch type disc brakes during the war under very arduous braking conditions, had shown that the disc brake concept had a far superior thermal performance than the drum brake, which could be exploited for use on road vehicles. As illustrated in *Fig. 9*, one of the first indications that some positive progress had been made in this direction was the publication of two articles in *The Motor* on the 1st and 8th October 1952, providing the general motoring public with a taste of things to come at the 1952 Earls Court Motor Show. The first article dealt with the design of the Dunlop and Girling spot type disc brakes, which were of very similar design, due to the Girling version being developed under Dunlop patents, based upon their very considerable experience in aircraft braking. As illustrated in *Fig. 10*, the second article dealt with the introduction by Lockheed of their spot type disc brake. It is of great historical interest that the two designs developed by the three manufacturers contained all the basic components of the modern disc brake, having remained unchanged over a period of forty five years.

These items as illustrated in *Figs. 9 & 10*, are :-

1. A top hat shaped cast iron brake disc fitted to the wheel stub axle.

2. A steel caliper with two sides which are bored out to accept hydraulically operated pistons, which is fixed to the vehicle by a steel torque plate, the hydraulic cylinders being connected by ports for the operating fluid.

3. A steel caliper bored out on one side only to accept hydraulically operated pistons, mounted at the extreme ends on steel pins which slide in the torque plate fastened to the vehicle.

4. Two pads of friction material mounted on steel plates which are located in the caliper and positioned either side of the brake disc.

The Dunlop and Girling design is made up of components 1,2, & 4 and is termed a fixed caliper disc brake, since the caliper which straddles the disc remains fixed to the torque plate with the friction pads being forced onto each side of the disc by means of the hydraulic pressure acting on the pistons.

The Lockheed design is made up of components 1, 3 & 4 and is termed a floating caliper disc brake since the caliper which straddles the disc is mounted on the axial steel pins which slide in the torque plate. When the hydraulic pressure acting on the pistons applies one pad to the disc, a reaction force is created which slides the caliper and pins and pulls the opposite pad onto the disc.

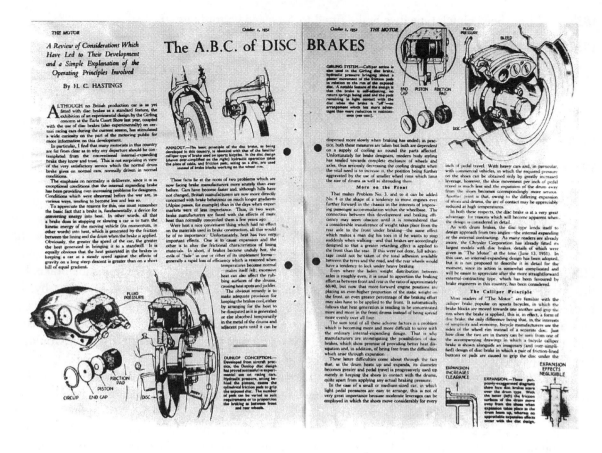

**Fig. 9 Details of the Dunlop and Girling car disc brakes in "The Motor" 1st October 1952.
(Courtesy "The Autocar")**

One of the most fundamental differences between the disc and drum brake is that the disc brake does not produce any self servo effect, the braking torque produced is entirely dependant upon the Coefficient of Friction of the friction material, the load applied to the pads by the hydraulic pistons and the effective working diameter of the brake disc. One advantage is that the disc brake is equally effective in both forward and reverse directions.

The disc brakes illustrated in *Figs. 9 & 10* were intended for use on the front brakes only, but in the event of fitting the disc brake to all four wheels consideration had been given to the provision of a separate handbrake. Girling produced a number of designs for a mechanical disc brake and Lockheed designed a hydraulic disc brake with one operating cylinder fitted with a mechanically operated push rod, and automatic adjustment for pad wear. It is of interest to note that the Dunlop and Girling design incorporated the Hydrastatic technique as used previously by Girling on drum brakes, with the disc pads rubbing lightly on the disc in order to keep fluid displacement and the associated pedal travel to the absolute minimum.

The Dunlop design had been subjected to trials on a Jaguar racing car in 1952 and had performed well at the sports car race at Rheims; and the same design was fitted to the BRM V16 racing car. In 1954, Dunlop disc brakes were fitted to the D Type Jaguar, the Austin Healey 100's, and in 1956, to the Jensen 541.

During 1953 Girling carried out a design reappraisal and in 1954 a production disc brake was fitted to single deck buses operated by the Birmingham and Midland Motor Omnibus Company, and production disc brakes were fitted in 1956 to the Triumph TR3, with the bridge part of the caliper removed to enable the inspection and easy removal of the disc brake pads without unbolting the caliper from the torque plate. This was the first quantity production car in the world to have disc brakes as standard equipment, with Ferodo disc pads being supplied as original equipment. Citroen

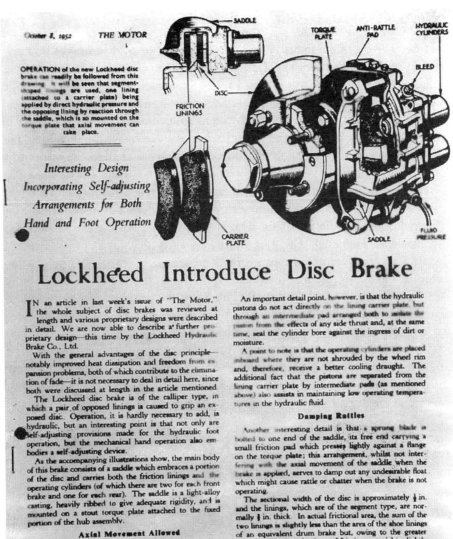

Fig. 10 Details of the Lockheed car disc brake in "The Motor" 8th October 1952. (Courtesy "The Autocar")

fitted their own design of sliding caliper disc brake to the DS19 model, which was well ahead of its time. The original Lockheed sliding caliper disc brake as announced in 1952 did not go into production, and in 1959 a fixed caliper disc brake was designed and fitted to the BMC 3 litre saloons and Alvis TD21 saloons, and in 1960 a new design was fitted to the MGA sports car.

As will be obvious from the foregoing dates, the disc brake after 1956 was gradually developing in its own right into an alternative braking system to the established drum brake, in the form of the fixed caliper system with progressive design refinements pertinent to each brake manufacturer. The superior performance of the disc brake had been positively established in 1952, as recorded in *The Motor* of 1st October which published details of comparative tests, carried out by Girling on a 100mph production car between drum brakes and disc brakes. The disc brakes had performed very well with light pedal effort and no fading of the disc pads, with an added bonus since the disc brakes were some 22% lighter than the drum brakes. But despite all the well founded optimism displayed by the brake makers, the motor vehicle industry was very reluctant to fit disc brakes on new production models. The ordinary motorist was very satisfied with both the performance and maintenance aspects of the drum brake, and there seemed no reason to introduce such a fundamental change. Even the brake industry and associated component suppliers were very hesitant, since the production facilities required to manufacture disc brakes were very different from those required for drum brakes. This aspect was not only in relation to machining but also foundry work, and the development of suitable friction materials and associated production facilities, all involving changes in working practice and considerable financial investment.

It was against such a background that the disc brake was gradually introduced on a production basis in the early 1960's, not only on the high performance models, but also on the small cars as a sales feature. The traditional association of the disc brake with the world of racing and the associated general public, which had been such a feature during its early development, was repeated in May 1964, when Donald Campbell created a new world land speed record of 403mph, with his turbine engine car Bluebird at Lake Eyre, South Australia. *Fig.11* illustrates the brake tests being carried out in the Ferodo test house, with Donald Campbell controlling the disc brake fitted to one of the brake testing dynamometers. The vehicle was fitted with two Girling sliding disc fixed caliper disc brakes on each wheel, each caliper having six specially made Ferodo disc brake pads, comprising a total of forty eight pads, capable of stopping the vehicle from 400mph in sixty seconds. With a maximum wheel speed of 2600rpm and a rubbing temperature at the surface of the friction material of 2200°F, the braking was even on all four wheels without any brake fade, proving beyond all doubt the superior thermal performance of the disc brake.

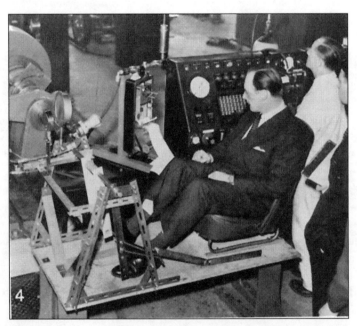

Fig. 11 Donald Campbell testing "The Bluebird" brake, Ferodo Test House, leading to the new land speed record of 403 mph in May 1964. (Courtesy Ferodo Ltd)

In the UK during the 1960's the disc brake scene was dominated by Dunlop, Girling and Lockheed with the eventual withdrawal by Dunlop from car brake manufacturing, and design and manufacture had also been undertaken by brake and vehicle manufacturers in Europe and America. Although the disc brake was obviously superior in many respects to the drum brake there remained a number of associated problems which greatly influenced its subsequent design. It was not always easy to fit a disc and caliper assembly into the space previously occupied by the drum brake, and this was only made possible by reducing the effective working diameter of the disc. The area of the disc pads was also considerably smaller than the area of the drum brake linings, with a consequent increase in the surface temperature of the pads, due to dissipating the same amount of energy over a smaller area. These two changes caused many problems for the friction material industry, and it was necessary to formulate new materials capable of withstanding the increased working pressure and temperature without loss of friction or excessive wear. A major problem with the disc brake was the provision of a suitable handbrake system when fitted to the rear wheels, which was either part of, or extra to the brake. In order to comply to the legal requirement it was necessary to provide an independently operated handbrake, and various designs were developed. These included a separate mechanically operated caliper, mounted adjacent to the hydraulic caliper, a small internal expanding brake operating within the bore of the disc mounting extension, and the installation of a self adjusting mechanically operated push rod system into the hydraulic caliper, which had the advantage of using the same running pads for parking purposes. One major problem associated with the use of the disc brake whilst in service was that of squeal generated during the brake operation. After various trials carried out by Lockheed, it was found that the squeal was greatly reduced if the friction pad was slightly offset to the centre line of the caliper piston, thereby altering the centre of pressure on the pad. This was achieved in a number of ways :-

1. A step was ground in the face of the caliper piston, which had to be reassembled in the same place relative to the caliper after removal for maintenance.

2. A thin shim was bonded on one half of the back of the friction pad backplate with an arrow to indicate direction of rotation.

3. The pad was supplied by the friction material supplier offset in relation to the centre of the backplate, which was shaped to prevent the friction pad being assembled incorrectly.

4. The friction material supplier inserted a thin layer of damping material on the backplate between the friction material and the backplate.

Fig.12 illustrates a Lockheed light duty fixed caliper showing the various component parts. The face of the piston has a shallow step to prevent brake squeal by altering the centre of pressure on the friction pad. *Fig.13* illustrates a Girling fixed caliper disc brake mounted on a vehicle, with details of the piston seal, which apart from preventing leakage of the pressurised hydraulic fluid, also due to its flexibility, ensures a constant running clearance of 0.004" (0.1mm) between the pad and the disc, after each brake application. This Hydrastatic technique was first used by Girling on the internal expanding drum brake and was included in the original Dunlop and Girling disc brake designs, eventually becoming standard disc brake practice.

Fig.14 illustrates a Dunlop mechanical handbrake attachment, which is self adjusting to allow for the wear of the friction pads, bolted to a Girling fixed caliper for rear wheel use.

Although the design of fixed caliper as shown in the illustrations functioned satisfactorily in service, its use did cause a number of problems :-

1. It was difficult to fit in the space previously occupied by the internal expanding drum brake.

2. The connecting ports for the hydraulic fluid passed through the bridge section above the periphery of the hot disc giving rise to an excessively high fluid temperature.

3. Producing the work piece and the machining of the bores and pistons was expensive.

4. It was bulky and heavy.

In an attempt to overcome these problems, both Girling and Lockheed introduced in the late 1960's, an entirely new concept, which eventually in the intervening years developed into the caliper system predominately in use on modern cars. *Fig.15* illustrates a Lockheed version of this design which is called the Single Cylinder Swinging Caliper. The caliper consists of a flat steel plate which is pivoted on the torque plate, with a hole machined in the middle to enable it to fit astride the brake disc. One cast cylinder and piston unit is fixed centrally to the plate and two friction pads retained by steel pins

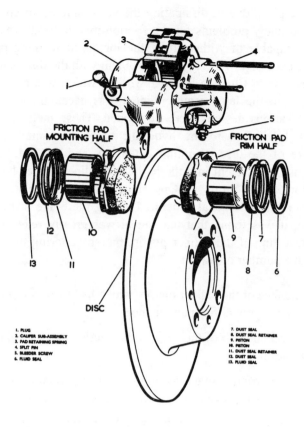

Fig. 12 Lockheed light duty disc brake (Courtesy A P Lockheed)

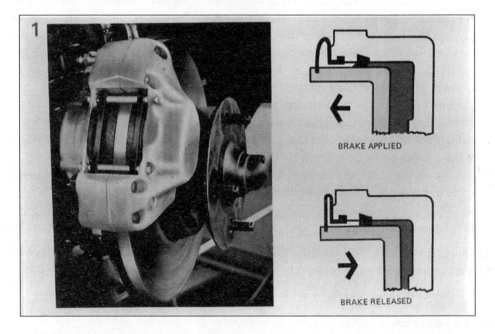

Fig. 13 Girling disc brake with seal details. (Courtesy Lucas Varity)

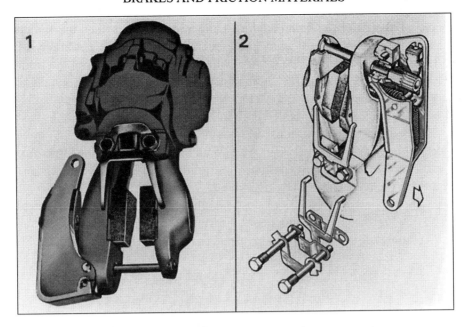

Fig. 14 Girling disc brake with Dunlop handbrake attachment. (Courtesy Lucas Varity)

Fig. 15 Lockheed single cylinder swinging caliper. (Courtesy A P Lockheed)

are mounted either side of the disc, one located in the plate adjacent to the piston, and the other located on a bracket fixed to the plate on the opposite side. When the cylinder is pressurised one pad is forced onto the disc causing a reaction force, which swings the plate on the pivot, thereby applying the opposite pad. To allow for the geometry of the swinging plate it was necessary to manufacture the pads in pairs, with the friction material tapered from end to end, the pads wearing parallel during use. Both designs could be used for rear wheel handbrake use with automatic pad wear adjustment.

Various designs of single piston sliding caliper disc brakes were produced in the UK, Europe and America in order to reduce cost and weight. *Fig.16* illustrates a Girling single cylinder two piston design with a yoke or plate which is free to slide in grooves on the side of the cylinder. One piston applies a friction pad directly on the disc and the other piston activates the yolk and applies the remaining friction pad to the opposite side of the disc. This caliper could also be fitted with mechanical operation, and automatic friction wear pad adjustment when used as a handbrake. An inherent advantage of the single cylinder sliding caliper design was that the cylinder could be positioned on the cooler inboard side of the vehicle, thereby reducing the possibility of overheating the brake fluid. It was also an ideal choice for use with the smaller wheels on the front wheel drive vehicles. In order to reduce the undesirable sliding friction associated with the sliding caliper design, especially when exposed to road dirt and water, the sliding action was eventually achieved by mounting the caliper on steel pins, with sealed built in lubrication in the caliper holes, as illustrated in *Fig.17*. By the 1980's this design in various versions had become very common, mounted with the operating cylinder on the inboard side of the vehicle. With the exception of the larger luxury vehicles, the general trend dictated by overall cost, has been to fit disc brakes on the front wheels and leading-trailing shoe brakes on the rear wheels. The intervening years to the present time have been devoted to refining the traditional designs in terms of cost, weight and low maintenance which is best dealt with by briefly reviewing the three components, bearing in mind that a great many of the older designs and configurations still exist, and in some instances remain standard equipment. See *Fig.12 & 13*.

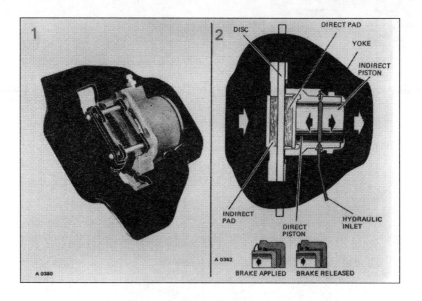

Fig. 16 Girling single cylinder two piston sliding caliper. (Courtesy Lucas Varity)

1. Caliper

When first introduced in the 1950's the caliper was machined from a heavily proportioned one piece steel unit, but eventually this developed into two half units which were bolted together with a slot in the bridge to allow access to the friction pads. This design facilitated easier machining of the cylinders, fluid ports and friction pad slots, the diameter and number of cylinders required depending upon the service level of the particular application involved.

The steel pistons were chromium plated for wear resistance, fitted with synthetic rubber fluid pressure seals and a seal positioned at the mouth of the cylinder to prevent ingress of dirt. This design was in general use until the development of the pin sliding caliper, the number of cylinders again depending upon the particular application. See *Fig.17*.

2. Disc

Basically the disc has remained unchanged, being manufactured in a high grade cast iron to a size and top hat shape to suit the particular application. The choice of cast iron is very important, since during the 1970's a serious braking phenomenon began to emerge, with vehicles dangerously drifting to the left or right during braking. After an intense period of investigation by Ferodo Research Division, this was identified as a problem associated with the amount of titanium, in the form of hard particles, that were present in the disc. If the amount of titanium was below 75 particles per mm² there was a marked increase in the Coefficient of Friction. Since it was impracticable to have matching discs the problem was eventually solved by joint liaison between Ferodo and the relevant foundries, ensuring the production of discs containing the correct amount of titanium in the castings.

A ventilated type of disc was used on the heavier American vehicles and this is now in general use due to its improved cooling performance compared with the solid disc. The ventilated disc consists of two flat discs, separated by air ducts, made as a one piece casting which is thicker than the ventilated disc.

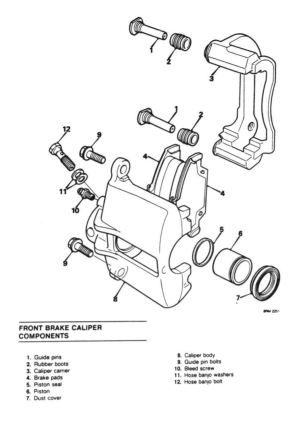

FRONT BRAKE CALIPER COMPONENTS

1. Guide pins	8. Caliper body
2. Rubber boots	9. Guide pin bolts
3. Caliper carrier	10. Bleed screw
4. Brake pads	11. Hose banjo washers
5. Piston seal	12. Hose banjo bolt
6. Piston	
7. Dust cover	

Fig. 17 Girling pin sliding caliper. (Courtesy Lucas Varity)

3. Friction Pad

This item commonly referred to as a disc pad has basically remained unchanged, consisting of a piece of friction material fixed to a steel backing plate, although there are a vast number of different sizes and shapes to suit the various calipers in use on UK and imported cars. Apart from the different friction material qualities which have been formulated since the 1950's, which is an ongoing process, the major change has been the method of securing the friction material to the backplate. The original riveting and bonding methods have been replaced with an integral moulding technique consisting of a number of holes drilled into the steel backplate, into which the friction material is forced under pressure during manufacture to form a number of spigots. The mating faces of the friction material and the backing plate are held together with a high temperature synthetic resin adhesive. If required, a radial groove is machined into the centre of the pad to assist in water deflection from the surface of the brake disc. This groove is also used on pads subjected to high temperature operating duties, to offset the differential expansion between steel and friction material, by making one pad into two smaller ones, thereby minimising bond failure.

One of the major problems associated with the production of the modern disc brake pad is the large number and variety of accessories which can be specified by the customer. These consist of pad retaining clips, anti-squeal or damping shims, an electronic sensing probe which is fitted into the pad to warn the driver when a pad change is imminent, and anti vibration rubber coating of the backplate. These items are not standardised and since there are a considerable number of different pad shapes and sizes involved, a good deal of preplanning is required, not only for production, but quality control procedures, packing and customer delivery.

Ref. (8), (9), (10), (11), (12), (13), (14), (15), (16), (17) & (18)

(2) Commercial Vehicle Disc Brakes

Although an attempt was made to introduce the disc brake into this market in 1956 with the fitting of Girling disc brakes onto single deck buses, as previously discussed, very little further progress as made. This was mainly due to the lack of suitable friction material, existing material suitable for consideration possessing a low Coefficient of Friction, requiring high application loads necessitating heavy calipers with a number of large diameter pistons. Even with medium weight vehicles it was necessary to have two calipers on each disc, which was costly and heavy and also gave rise to thermal problems due to the reduction of the disc cooling area. The increase in cost and weight and the much smaller market compared to the private car, inhibited the development of the commercial vehicle disc brake for some years, but eventually in the 1960's it was fitted to some of the lighter delivery vans, being a logical development from the private car. With the development of a more suitable friction material a design was introduced by Deutsche Perrot Bremse, and eventually large pin sliding calipers and double disc brakes were developed in Germany and Italy using ventilated discs to aid heat dissipation. The long hauls and steep Alpine descents associated with road transport in Europe, encouraged the use of the disc brake on heavy commercial vehicles, whilst in the UK, with less adverse operating conditions, adequate braking was provided by increasing the width of the drum and brake linings. Heavy commercial vehicle disc brakes have been developed in America, development having started in 1965 with Rockwell International starting production in 1981, being now the sole supplier of air disc brakes for trucks, tractors and trailer units, Girling have produced designs in the UK, and with the great increase in UK heavy goods traffic to and from Europe, via the channel ferries and tunnel this development will continue. *Fig.18* illustrates a Girling heavy commercial vehicle mechanically operated disc brake. To obtain maximum utilisation most of the vehicles used for this type of haulage are of the tractor and trailer type, with disc brakes fitted to the front wheels of the tractor and internal expanding drum brakes fitted to the rear wheels and trailer.

Ref. (20)

Reaction Beam Air Disc Brake

APPLICATION
Front or rear brake with parking typically for 12.0 Tonne to 19.0 Tonne GVW trucks and buses.

CLAMP LOAD AT 60%g
Up to 116.0 kN (26,080 lbf).

FEATURES
- Air operation
- Uniform lining pressure
- Separate clamping and lining drag force
- Radial lining removal
- Identical inner and outer linings
- Low parasitic drag through sealed slides
- Mechanical parking
- Automatic adjustment

OPERATION
A single mechanical actuator, comprising two rotary helical ramps with a ball race interposed between the two, generates the clamp force from a lever and conventional air actuator. This assembly is located within the inboard beam of the clamping frame and pressurises the inboard lining directly onto the rotor. Outer lining pressure is achieved by an equal reaction force through the clamping frame.

Fig. 18. Girling commercial vehicle disc brake. (Courtesy Lucas Varity)

(3) Motor Cycle Disc Brakes

The braking system fitted to motor cycles from the 1930's was the cable operated mechanical internal expanding drum brake. This was fitted to each wheel being of the leading-trailing design, although some of the racing models developed in the early 1950's fitted a cable operated mechanical two-leading shoe brake, with each cam connected externally by a linkage. The first successful disc brake used on a motor cycle was in 1966, when a Lockheed fixed light alloy caliper design was fitted to a racing cycle, and eventually this type of caliper was fitted to most Norton and Triumph production models. As was the case with the car, this type of caliper was eventually replaced with the single cylinder pin sliding caliper, which due to its smaller dimensions and light weight construction was an ideal choice for motor cycles, being extensively produced in Japan under licence from Girling. The current situation remains unchanged with one disc brake unit fitted to each wheel, or on the heavier models two disc brakes fitted either side of the front wheel and one on the rear wheel. There is now a tendency to fit calipers with a number of small cylinders.

The discs on motor cycles are much thinner than car discs, being manufactured from stainless steel in order to provide strength and lightness. Since a good deal of water is picked up from the road surface in wet conditions, some discs contain holes or slots which enable the disc pads to function without a film of water on the disc surface.

Ref. (21)

(4) Tractor Disc Brakes

A brief reference to this particular vehicle is of interest since unlike the car, the tractor is used in somewhat hostile conditions for long periods of time without regular maintenance. In order to satisfy these conditions a number of brake designs have been used which in chronological order are :-

1. A mechanically operated duo-servo shoe brake fitted to each wheel with independent operation.

2. A hydraulically operated Dunlop patented disc brake introduced by Girling with piston dust covers to obviate contamination by mud and dust.

3. A Girling clutch type disc brake, operated mechanically from a footpedal, working in a cast iron housing on each wheel, using organic friction facings. This unit is termed the Girling Multi Plate Disc Brake.

4. Another version of the Girling Multi Plate Disc Brake using sintered metal friction facings totally immersed in lubricating oil from the gearbox, which removes the heat from the surface of the facings, thereby minimising wear and replacement. This unit is termed the Girling Oil Immersed Multi Plate Disc Brake.

Ref. (22)

Vehicle Actuating Systems

(1) Mechanical

Private Car

The band brake, which was one of the earliest brakes to be used on cars, was operated from the hand lever or foot pedal by means of steel rods and levers pivoted on cross shafts. When the leading-trailing shoe brake was introduced, a similar system was used to operate the brake cam, much of which was duplicated when in 1904 legislation came into force, stipulating that road vehicles must have two independent brakes. Apart from the maintenance problems associated with the effect of road dirt, mud and water on the various pivots and associated components, it was very difficult to ensure even braking due to the accumulated wear of the pivots and bearings, giving rise to a good deal of slack and loss of pedal travel. This situation was greatly improved by Captain A.H.Girling when he replaced the rotary cam in the leading-trailing shoe brake with the cone expander, which required a straight pull to operate instead of a rotary motion. This system did not require the duplicated pull rods and cross shafts, being replaced with one central rod, pivoted levers and cross rods to the four brakes, with one foot pedal operating all four brakes, and a hand lever independently operating the rear brakes. The system operated under tension and since the number of pivots was considerably reduced, provided a tight system without slack, which greatly improved both braking and pedal movement. Girling also introduced a hydro-mechanical system with hydraulic actuation for the front brakes, and mechanical actuation for the rear brakes. This was a very neat way of overcoming the problems associated with the use of mechanical actuation on the steerable front wheels.

As previously discussed the leading-trailing shoe brake did provide a self servo effect, and for average sized vehicle weights this was sufficient to provide braking with reasonable pedal effort. With the heavier vehicles this was not the case and despite the leverage provided by the foot pedal, a great deal of leg effort was still required from the driver in order to brake the vehicle. In an attempt to overcome this problem, Hispano Suiza fitted a mechanically operated servo on their 1919 model, consisting of a drum and band brake fitted to the output side of the gearbox, with one end of the band being fixed to the gearbox and the other end being connected to the brake operating rods. When the band brake was operated, the resultant force from the friction material tightening around the drum provided extra effort to that provided by the foot pedal, being a rather unique case of one brake operating another. Other designs of servo unit were used based upon plate clutches or the use of vacuum from the engine induction manifold, as used in the Dewandre vacuum servo unit.
Ref. (23)

Commercial Vehicle

The drum brakes for commercial vehicles were usually made by the manufacturer, being of a leading-trailing design activated by pull rods and lever; as developed for use on the private car. Girling introduced in the 1930's a two-leading shoe brake which was mechanically operated.

(2) Hydraulic and Pneumatic

Private Car

Although the hydraulic brake actuation system used on the car has in principle remained the same since its introduction in the 1920's, its initial development depended to a great extent upon the availability of reliable flexible pipes, joints and cylinder seals. During the 1939-1945 war, a great deal of effort went into the development of synthetic rubber components for use on aircraft and military vehicles, and in the immediate post war years, this technology was used to manufacture similar components for use with vehicle hydraulic brake systems. The most important part of the system is termed the master cylinder which replaces the mechanical foot pedal lever, and provides hydraulic pressure through thin walled steel tubing and flexible hoses to the front and rear brakes. A Girling centre valve master cylinder is illustrated in *Fig.18* and consists of a combined reservoir and cylinder unit containing a piston which is connected to the brake foot pedal. The hydraulic fluid from the reservoir enters the cylinder through an inlet situated at the opposite end to the piston, and the hydraulic fluid to the brakes flows out through an outlet in the top of the cylinder body into the brake pipes. When the foot pedal is depressed the piston moves forward in the cylinder, and by means of a plunger closes a valve seal to the reservoir, and forces pressurised fluid throughout the outlet to the brakes. When the foot pedal is released a spring in the cylinder forces the plunger and piston back, releasing the hydraulic pressure, and opening the valve seal to the reservoir and atmosphere. If any loss of hydraulic fluid does occur this is automatically replaced from the reservoir. This sequence of events can be followed in the two diagrams in *Fig.19*.

A number of different designs of master cylinder have been introduced by Girling and Lockheed, since the introduction of the hydraulic two-leading shoe brake in 1947. The essential function remains the same, to consistently provide hydraulic pressure, which can be as high as 1500lbs per square inch (100 bars), under all driving conditions.

The inclusion of air in the form of bubbles in the hydraulic system is very undesirable, since this allows compression to take place with consequent increase in pedal movement and partial or complete loss of braking. This dangerous situation has been overcome by fitting bleed nipples into the brake operating cylinders, which are situated at the highest point in the brake system, as illustrated in *Figs.12 & 17*. The bleed nipple is a very simple but effective device consisting of a hexagon head and screwed body with a taper seating, the hexagon head and the body is centrally drilled to a cross port drilled above the taper seat. It is screwed into a hole drilled in the cylinder which is opened out for a taper seat and internal thread. When the nipple is screwed into the hole and tightened, the taper seats seal the system, when partially released the pressurised fluid and air bubbles are forced up into the nipple cross port and ventilate to atmosphere through the hole in the body and hexagon head.

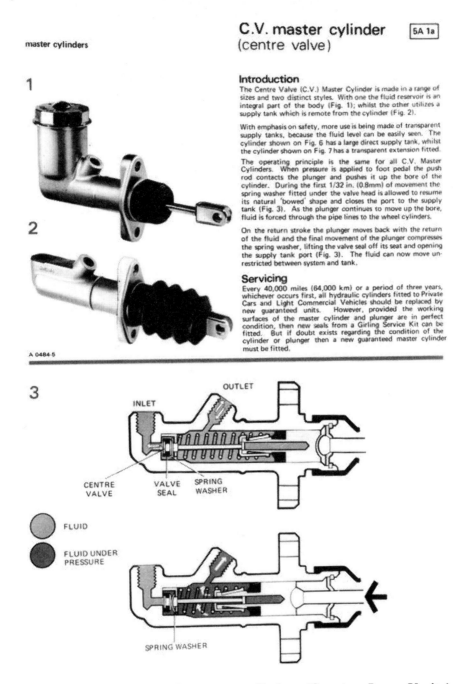

master cylinders

C.V. master cylinder [5A 1a]
(centre valve)

Introduction

The Centre Valve (C.V.) Master Cylinder is made in a range of sizes and two distinct styles. With one the fluid reservoir is an integral part of the body (Fig. 1); whilst the other utilizes a supply tank which is remote from the cylinder (Fig. 2).

With emphasis on safety, more use is being made of transparent supply tanks, because the fluid level can be easily seen. The cylinder shown on Fig. 6 has a large direct supply tank, whilst the cylinder shown on Fig. 7 has a transparent extension fitted.

The operating principle is the same for all C.V. Master Cylinders. When pressure is applied to foot pedal the push rod contacts the plunger and pushes it up the bore of the cylinder. During the first 1/32 in. (0.8mm) of movement the spring washer fitted under the valve head is allowed to resume its natural 'bowed' shape and closes the port to the supply tank (Fig. 3). As the plunger continues to move up the bore, fluid is forced through the pipe lines to the wheel cylinders.

On the return stroke the plunger moves back with the return of the fluid and the final movement of the plunger compresses the spring washer, lifting the valve seal off its seat and opening the supply tank port (Fig. 3). The fluid can now move unrestricted between system and tank.

Servicing

Every 40,000 miles (64,000 km) or a period of three years, whichever occurs first, all hydraulic cylinders fitted to Private Cars and Light Commercial Vehicles should be replaced by new guaranteed units. However, provided the working surfaces of the master cylinder and plunger are in perfect condition, then new seals from a Girling Service Kit can be fitted. But if doubt exists regarding the condition of the cylinder or plunger then a new guaranteed master cylinder must be fitted.

Fig. 19 Girling centre valve master cylinder. (Courtesy Lucas Varity)

As with the mechanical drum brakes it was still necessary to provide servo assistance, especially with the disc brake which does not produce any self servo effect. This was introduced by Lockheed and Girling in 1954/1955, and a Lockheed version is illustrated in *Fig.20*. The one piece unit is connected into the hydraulic circuit between the master cylinder and the brakes, and consists of a hydraulic slave cylinder with a control valve, connected to a pressed steel chamber and booster piston, the rod of which is connected to a piston in the slave cylinder. When the master cylinder is operated, the hydraulic pressure operates the control valve and allows a vacuum sourced from the

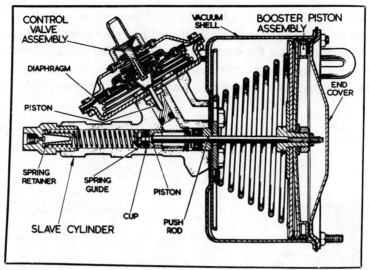

Fig. 20 Lockheed vacuum servo unit. (Courtesy A P Lockheed)

engine induction manifold to be formed in the chamber, creating a pressure differential across the booster piston thereby moving the rod and piston in the slave cylinder, which boosts the existing pressure thus assisting the pedal effort. Eventually the piston was replaced by a diaphragm and the chamber was maintained under vacuum, the slave cylinder being pressurised when the atmosphere was allowed into the chamber by the control valve. This unit known as the Vacuum Suspension Servo provided a much quicker response than the previous system. In order to safeguard against loss of vacuum, a reservoir and non-return valve were fitted into the brake pipe circuit between the engine induction manifold and the servo unit. An alternative system has been developed which has now become standard practice on cars, called the Direct Acting Servo. The unit is similar to the vacuum suspension unit, but is fitted between the brake pedal and the master cylinder with servo assistance being applied directly to the master cylinder push rod. Since this unit is not part of the hydraulic circuit it is far superior in terms of operation and maintenance than the previous designs. Because it is not possible to obtain vacuum from the diesel engine, which is now used extensively for cars, it has been necessary to provide a vacuum pump driven from the engine. *Fig.21* illustrates the Lockheed actuation systems for medium, light and heavy duty brake systems, showing the various components including the pressure regulating valve. This was developed to enable maximum braking effort to be used without dangerous rear wheel locking for normal conditions of adhesion. The valve ensures that pressure to the rear brakes is kept to a certain set value, with any additional rise in pressure being applied to the front wheel brakes, thereby allowing for weight transfer.

With the vast increase in the car population and the higher speeds attainable on the motorways, much greater emphasis is being placed on safety by the car industry, and a number of items have been included in the actuation systems to enable safer braking. The development of the brake warning light system on the mechanically actuated vehicles was a valuable aid to safety. The same basic principle is used with hydraulic brakes, by fitting an electrical pressure sensing switch into the hydraulic circuit, which illuminates the lights on the rear of the vehicle when the brakes are applied. Although component failures in car braking systems are rare, a good deal of attention has been given to either splitting or duplicating the brake circuit, to enhance the safety of the braking system. This trend of events started in America when in the 1960's vehicles had to be fitted with split systems. In the UK this development took many forms from tandem master cylinders providing pressure to two separate circuits, in various combinations of front and rear brakes, to actual duplication of the system. The current situation is a compromise between cost and performance and the system termed diagonal split is used. This comprises a tandem master cylinder, which pressurises two separate circuits to the left hand front brake with the right hand rear brake, and the right hand front brake with the left hand rear brake. In the event of failure of one of these circuits, the remaining brakes will stop the car without serious rear wheel skidding.

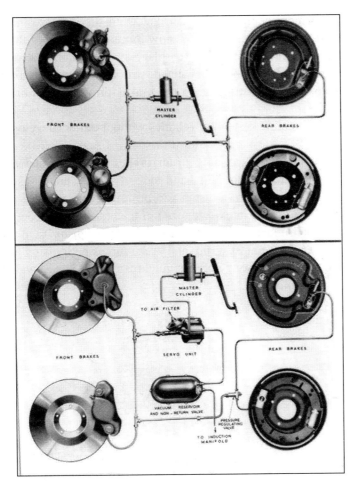

Fig. 21. Lockheed light and heavy duty drum and disc brake systems. (Courtesy A P Lockheed)

Equipment to prevent the skidding of cars has also been the subject of a good deal of development for many years. This dates back to the work carried out in the 1950's by Dunlop to prevent the locking of aircraft undercarriage wheels during landing, called the Dunlop Maxaret System. A good many anti-lock systems, as they are termed, have been developed but the associated cost prevented their acceptance on a production scale, apart from the luxury cars. In view of the more stringent attitude towards road safety this view has changed and various types of anti-lock systems are now much more common. Apart from the degree of electronic sophistication, the basic concept is the same and consists of a slip detection device fitted to the wheels, which send out signals to an electronic valve system for reducing and controlling the hydraulic pressure to the brakes.
Ref. (24), (25), (26), (27)& (28)

Commercial Vehicle
Hydraulically activated commercial vehicle brakes were introduced by Lockheed during the 1930's. The actuation systems used on vehicles with a gross weight of up to 12 tons, which covers the light to medium commercial vehicle range including passenger service vehicles, are similar to those used on private cars. The hydraulic system and servo units are suitably sized up for brakes of more robust proportions, to cope with the arduous duties associated with delivery vehicles. In order to assist with the heavier foot pedal loads, increased hydraulic pressure has been provided by the use of air pressure servo units, or hydraulic servo units and hydraulic fluid pressure accumulators.

The actuating system used on vehicles in excess of 12 tons gross vehicle weight, which are termed heavy commercial vehicles, is entirely different due to the increased size of brake components and circuits. The use of hydraulic actuation is inhibited due to the amount of displacement involved, necessitating the use of compressed air. This was introduced by the Westinghouse Brake Company in

1922 and the Clayton Dewandre company in 1934, and with the increase in lorry traffic after the 1950's especially in recent years, pneumatic brakes now form a major part of the brake industry.

Although a great deal of development has taken place since pneumatically operated brakes were introduced, the basic concept has remained unchanged, with a compressor supplying compressed air to a storage reservoir. This is then distributed to the brakes through pipework and a foot operated valve in the driving cab.

The development of the actuating system on vehicles without trailers developed as follows :-

1. In order to operate the internal expanding drum brakes a unit termed a brake chamber was developed. This item is illustrated on the left hand side of *Fig.22* and consists of a pressed steel chamber, containing a rubber diaphragm, a piston and push rod with a return spring. The end of the push rod is connected to the cam lever or expander rod of the brake. When compressed air is admitted to the rear of the chamber, the rubber diaphragm is forced onto the piston face and the rod applies the brake. When the air is released the return spring and brake forces both piston and diaphragm back to the original position. In order to release the brakes quickly, a quick release valve or as it is termed a dump valve, is fitted into the pipework, which evacuates the compressed air from the system faster than the drivers foot valve. The rear wheel brakes were also used for parking, being operated independently by a cable and hand lever from the cab. This one line system is not fail safe, since the brakes would be inoperative if the air pressure was lost, with the exception of the handbrake.

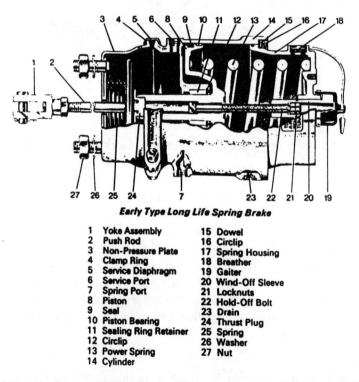

Early Type Long Life Spring Brake

1	Yoke Assembly	15	Dowel
2	Push Rod	16	Circlip
3	Non-Pressure Plate	17	Spring Housing
4	Clamp Ring	18	Breather
5	Service Diaphragm	19	Gaiter
6	Service Port	20	Wind-Off Sleeve
7	Spring Port	21	Locknuts
8	Piston	22	Hold-Off Bolt
9	Seal	23	Drain
10	Piston Bearing	24	Thrust Plug
11	Sealing Ring Retainer	25	Spring
12	Circlip	26	Washer
13	Power Spring	27	Nut
14	Cylinder		

Fig. 22 Clayton Dewandre spring brake unit. (Courtesy WABCO Automotive UK Ltd)

2. In order to safeguard against this possibility a system was developed with two separate circuits. The reservoir was divided into two chambers with separate pipework for front and rear brakes, both circuits being controlled through a dual foot valve in the cab. The rear wheel brakes were still used for parking, but the manual application from the cab hand lever was assisted by compressed air.

3. On 1st January 1968 the Ministry of Transport introduced new standards concerning certain aspects of commercial vehicles including braking. It was required to provide parking brakes on all vehicle wheels, and this was eventually achieved by using an item called the spring brake unit, which had been used in America for emergency purposes. This unit is illustrated in *Fig.22* and consists of a normal brake chamber with an extended body, which contains a piston and a very powerful

compression spring. Two separate reservoirs are used, one being used to supply compressed air to the brake chamber through the drivers foot valve for routine braking. The other reservoir being used to supply air to the spring chamber through a hand regulating valve in the cab, which during normal running forces the piston and spring into the compressed position. When the vehicle requires parking and the brake chamber is out of use, the compressed air acting on the piston is released by the hand regulating valve. This allows the spring to expand forcing the piston locating boss onto the back of the brake chamber diaphragm, which applies the brake through the rod. This system is also fail safe since in the event of loss of air in the service braking line, the driver can evacuate the spring chamber and apply the brakes, the deceleration being dependant upon the rate at which the compressed air is released through the hand regulating valve.

In order to comply with European Common Market recommendations further legislation was introduced in January 1975. This required the duplication of the service braking line with a dual foot valve, and a supply reservoir which was divided into two chambers. The emergency and parking brake line remained unchanged.

Although the basic braking requirements as described remain unchanged, further European Common Market legislation has been introduced for single vehicles, draw-bar trailers, and tractor trailer units which are now in wide use with three line braking.

In order to take advantage of available road adhesion, axle load sensors have been developed, which automatically adjust the brake operating pressure to match the load being carried. Automatic slack adjusters were introduced in 1971 ensuring compensation for brake lining wear, and in 1978 the introduction of plastic piping reduced the risk of failure of the actuating circuit by corrosion.

Anti-skid systems have been developed, especially for use with tractor trailer units, in order to avoid the highly dangerous jack knifing associated with this kind of vehicle, when directional control is lost, causing the tractor and the trailer to swing towards each other.

Ref. (29), (30), (31), (32), (33), (34), (35) (36) & (37)

Motor Cycle Brakes
The hydraulically operated sliding pin caliper disc brakes fitted to motor cycles, are actuated by small special purpose master cylinders, the front brakes being operated by hand lever and the rear brakes by foot pedal.

Ref. (38)

Brake Fluid
Brake fluid plays a very important role in the actuating circuit for hydraulic brakes, and has been an integral component dating back to the first hydraulically operated drum brakes introduced during the 1920's. The basic requirements of brake fluid have remained unchanged, it must apply the brakes without damaging any of the associated components, and must not be affected by operating temperatures and climatic changes. Brake fluid was developed by Charles Wakefield, who had created C.C.Wakefield and Company in 1899 to produce lubricating oils for the railway industry. In 1909 he extended this activity to the private car by producing a blend of oil which contained castor oil, which was marketed as Castrol, a name which has become inseparably linked with the motor industry. He took out a patent for car brake fluid in 1926 based upon castor oil and alcohol, which he claimed was far superior to existing products in relation to its effect upon rubber and leather seals and packings. His company became the accepted leader in the field, and has since formulated blends capable of working under very wide variations of temperature, either due to climate or usage. Brake fluids are specified on an international basis to comply with the American Society of Automotive Engineers (SAE) and American Department of Transportation (DOT).

Ref. (39) & (40)

Railway Brakes
As discussed in the introduction, up to the period of the 1950's, very little change had taken place in railway braking from the previous years, this was still carried out by applying cast iron blocks to the periphery of the wheel by means of vacuum, compressed air, steam or even by hand. A good deal of work had been carried out before the 1939-45 war in France, Germany and America, to provide high

speed braking on the 100mph steam powered passenger trains, by modifying the pneumatic system. Similar work was carried out by the LMS and LNER with the vacuum brake, culminating in the 126mph world speed record by Mallard in 1938 during high speed braking trials. In the UK carriage stock continuous brakes were applied by vacuum, locomotive brakes were applied by steam pressure, and the vast majority of freight wagons had non continuous hand applied brakes. These were made up into freight trains with a guards van at the rear fitted with hand wheel operated brakes for braking and parking, the brakes being applied at the top of a descent and allowed to rub continually until released at the bottom. Some effort was made in the mid 1950's to improve the braking of freight trains and a programme of work was drawn up to fit vacuum operated continuous brakes, but in general the existing actuation systems using cast iron blocks were quite adequate for both passenger and freight duties. Even if a change to a different system had been justified, the immense task and the financial implications of modifying thousands of passenger and freight vehicles would have been prohibitive. The continuous vacuum brake system was a well proven fail safe system that had become part of railway working practice over a period of some forty years with a good safety record. This was also the case with the hand braked freight wagons which moved at a relatively slow rate between sidings, where brake adjustments could be made to suit the various gradients with relatively few accidents.

Unfortunately the exception to this was the serious accident which happened at Chapel-en-le-Frith on the 9th February 1957 on the London Midland Region Manchester to Buxton line, as illustrated in *Fig.23*. The sole cause of the accident was loss of braking due to the failure of a brake valve steam

Fig. 23 Collision at Chapel en le Frith South Station, 9th February 1957. (Courtesy "The Buxton Advertiser")

pipe joint, on the locomotive of a freight train being banker assisted up the incline out of Buxton. The train was pushed over the summit of the incline but was unable to stop, and since the fireman could not manually apply the brakes on the moving wagons, it proceeded at 60mph down the incline to Chapel-en-le-Frith South Station, where it caught up and collided with the guards van of a preceding freight train. In the ensuing impact both the driver of the locomotive and the guard in the van were killed, and the two freight trains and station premises sustained considerable damage. In the report on the collision by the Ministry of Transport and Civil Aviation published on the 10th July 1957, Brigadier C.A. Langley concluded that if the freight train had been fitted with the continuous vacuum brake the accident would not have occurred. He went on to state that it was the intention of the British Transport Commission to equip all freight rolling stock, and that during 1957-58 over 200,000 new or rebuilt freight wagons would be fitted with continuous brakes. This serious accident became national news, and John Axon the driver of the locomotive was posthumously awarded the George Cross for devotion to duty, fortitude and courage in highly dangerous and alarming conditions, as reported in the London Gazette of 7th May 1957.

It was ironic that this accident should have occurred at Chapel-en-le-Frith, which as dealt with in Part One is the home of Ferodo Brake Linings, where in 1954 a development programme was started to provide disc pads for a prototype railway disc brake produced by Girling. This work was being carried out by Girling at the request of British Rail who wanted to reduce the cost associated with the frequent replacement of cast iron blocks. Although some work had been carried out before the 1939-45 war by Knorr Bremse in Germany and Budd in America with pneumatically operated disc brakes, this request marks a milestone in the history of railway brakes. Railway brake designers were aware that the block brake was not suitable for speeds above 100mph, and for higher speeds it would be essential to use some form of disc brake. Apart from the increase in speed there were also a number of other advantages :-

1. By removing the braking from the wheel periphery it became possible to have a dedicated disc for braking purposes only.

2. The disc could be made of cast iron which was the traditional material used for braking purposes.

3. The disc would not be contaminated with grease and dirt from the rails.

4. The thermal performance of the disc brake is far superior to the block brake.

5. There was far greater scope to provide a suitable friction material.

6. The running profile of the wheel periphery was not damaged or disfigured.

The first running trials were carried out by Girling in liaison with BR in 1955, using a two car electric unit travelling at 68mph on suburban services in the London area with pneumatic brake operation. The unit consisted of a motor and trailer coach, each unit having two four wheel bogies, the leading bogey of the motor coach having two axle hung nose suspended motors, with the remaining axles being non motored.

The disc brake caliper, as illustrated in *Fig.24* comprised two forged steel arms or levers, centrally pivoted on a steel bracket bolted to the bogie, which straddled a cast iron disc, the inner end of each arm being fitted with a vertically pivoted holder containing a friction pad. In between the outer ends of the arms was positioned a hydraulic piston and cylinder expander unit operated from an air-hydraulic master cylinder unit bolted to the bogie frame, which when operated expanded the outer ends of the arms, forcing the friction pads onto each side of the disc. The non motored bogies were fitted with ventilated cast iron discs, mounted on steel hubs which were shrunk onto the axle inboard of each wheel. This arrangement was not possible on the powered bogie, due to the space limitation imposed by the traction motor, and cast iron discs mounted on steel centres were bolted on each side of the wheel web, the discs having cooling fins cast on the inner side. Comparison trials were carried out between the disc braked unit and a standard block brake unit. The results were very encouraging both in terms of brake performance and the confidence of the drivers, who initially were suspicious of the retardation feel compared to cast iron blocks. The good results justified further trials and these were carried out on twelve cars fitted with pneumatically operated disc brakes. Trials were also carried out on freight hopper wagons, where due to space limitation it was difficult to fit the brake rigging required for block brakes. The disc brakes were actuated by vacuum and the brake pads clamped onto disc rubbing surfaces machined in the steel wheel webs.

Fig. 24 Girling prototype railway disc brake. (Courtesy SAB Wabco BSI)

The initial programme of work started in 1955 was completed in 1959 and a great deal of service experience was obtained by Girling with disc brakes fitted to Freightliner vehicles, and AM10 four car multiple units commissioned in 1966 for use at 75mph on the newly electrified London to Manchester line. The overall situation was also greatly improved by the decision made by British Rail during the mid 1960's to change from vacuum to pneumatic actuation enabling the use of much smaller caliper operating cylinders. It is of historical interest to note in this respect, as mentioned earlier in Part One, that George Westinghouse visited the UK in 1871 and again in 1878, in order to exploit his pneumatically operated block braking system, without success.

The design modifications and improvements carried out during these years form the basis of the modern railway disc brake and can be categorised as follows :-

1. Due to the pad holders being pivoted vertically to the centre line of the disc taper pad wear occurred across the pad width. This was overcome by restraining the pivoting action with a swinging tension link, which was connected to the top of each holder and suspended from the underside of the vehicle with hinges. The links also removed the torque reaction from the caliper arms.

2. An automatic wear adjuster mechanism was designed and fitted to obviate excessive pad clearance and reduce air consumption.

3. Pneumatic vehicle type air brake chambers were used to actuate the caliper.

4. The use of machined rubbing surfaces on the wheel web termed integral discs proved to be unsatisfactory, and eventually both this technique and the steel centred disc were replaced with cast iron plates termed cheek plates. These plates had cast cooling fins on the rear side and were split into two halves for easy fitting and removal. Each plate was fixed to the wheel with bolts, the holes being

accurately drilled in the web with sufficient clearance to accept thermal expansion and contraction of the plates under operating conditions.

5. To facilitate easy removal of the axle mounted discs the steel hub was suitably drilled for oil injection equipment.

6. The disc pad was grooved into a number of segments to ensure even distribution of the applied load, and also to improve cooling and disposal of wear products.

7. The caliper arm fixed pivot system was redesigned utilising a yoke system, which allowed the pad on the piston rod side to be applied first, giving rise to a reaction force which applied the pad on the opposite side. This aspect is of interest since Girling used the same principle with their version of the swinging arm car disc brake caliper.

The trial running also gave rise to two rather important functional aspects, one with far reaching effects :-

1. Although undesirable, it was practical to run trains formed of both block brake and disc brake stock. This aspect was very important in the 1960's and 1970's when most existing stock was fitted with block brakes.

2. As was evident from the first trials carried out in 1955 there was a reduction in rail/tyre adhesion when disc brakes were fitted, since the cleansing action of the block on the wheel tyre periphery was removed.

An attempt was made to overcome this problem by fitting wheel slide detection equipment, consisting of an alternator mounted at the end of the axle, which sensed any wheel lock up, and through an electronically processed signal actuated an electro pneumatic dump valve to reduce the operating pressure. With the gradual phasing out of the heavier locomotive hauled passenger trains, and the gradual introduction of the lighter Electric and Diesel Multiple Units the problem of reduced rail/tyre adhesion has become serious. During the Autumn when the rails are coated with a film of vegetable matter from the falling leaves, loss of adhesion has led to train cancellation and consequent disruption of the working timetable. Although this problem has also occurred on units with brake blocks the introduction of disc brakes has been a major contributory factor.

In order to alleviate this problem when Girling cheek plate disc brakes were designed for use on the Mark III coaches of the prestigious 125mph High Speed Trains in 1970, as illustrated in *Fig.25*, a cast iron block, termed a scrubber block, was also installed, which rubbed on the tyre periphery when the disc brakes were applied. Apart from cleaning the wheel tyre profile, this block was also used for train parking purposes. Scrubber blocks have also been used on the Class 158 and 159 Diesel Multiple Units which are in use over various routes where contamination has insulated the wheels causing track circuit problems. It is of historical interest to note the combined use of the best features of block and disc brake in this application, as was the case with the use of front disc brakes and rear drum brakes on road vehicles.

Although the development of the railway disc brake has been dealt with in terms of the Girling system on British Rail, a great deal of contemporary development was also taking place on the Continent and in America, and it is now fitted in various forms to rail vehicles throughout the world. One of the most outstanding applications has been that of the 190mph TGV and Channel Tunnel Eurostar coaches with four disc brake units per axle using sintered metal friction pads and steel discs, being a fundamental change to the traditional organic friction material cast iron relationship. Another interesting application is the use of a shaft mounted disc brake, on the drive system of the Class 91 locomotives used on the East Coast Route, which is extra to the brake blocks.

As was the case with the car disc brake space limitations can create problems for the brake supplier and invariably the system has to be designed to suit each particular application. An interesting alternative which is available for use in such circumstances, depending upon the speeds required, is the system developed by SAB some years ago, as illustrated in *Fig.26*, consisting of a

Disc brake pads for trains

TWENTY YEARS' EXPERIENCE

Herbert Frood commenced his experiments with friction materials in 1897, his attention at first being given to brake blocks for horse drawn vehicles; he soon widened the scope for his products, however, to include motor vehicles and railway rolling stock. At that time, underground railways in particular were interested in obviating the accumulation of metallic dust from cast iron blocks; Ferodo blocks, which were non-metallic, enabled this requirement to be met and in 1907 orders were obtained for the London system, the Paris Metro soon following this lead.

Since those days, Ferodo moulded brake blocks have superseded the fabric types and have entered service also on surface railways; an extensive market for these now exists all over the world, the composition block competing with the traditional cast iron which is still widely used.

Railway engineers are, however, gradually changing their allegiance from the block brake, which operates against the wheel tread, to the disc brake, particularly for use on the faster trains which are coming into service. Railway disc brakes have always made use of moulded friction materials—similar to car disc brake pads, but much larger (1)—and Ferodo Ltd has developed materials specially suited to their requirements; the company is, of course, well equipped for the testing of such materials, having an inertia test machine large enough to mount a full size unit, as described in issue 1 of *FITN*.

The development of railway disc brakes started in the UK in 1954 when Ferodo Ltd and Girling Ltd began a close collaboration to produce equipment for British Rail. Railway engineers are conservative by nature and the process of developing and proving a product which meets their very stringent requirements with regard to safety, performance and durability has been a long one. An example of only one of the many special considerations which have had to be kept in mind relates to adhesion, which is similarly variable when using steel tyres and rails as when running rubber tyres on an assortment of road surfaces; with four axles per coach and the driver unable to tell by his own senses if wheel sliding is occurring, a refined system of control is needed if flatting of wheels is to be avoided. Adhesion on rail seldom exceeds 0.2, a much lower value than road vehicles can normally achieve.

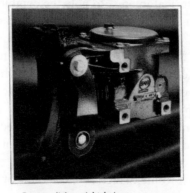

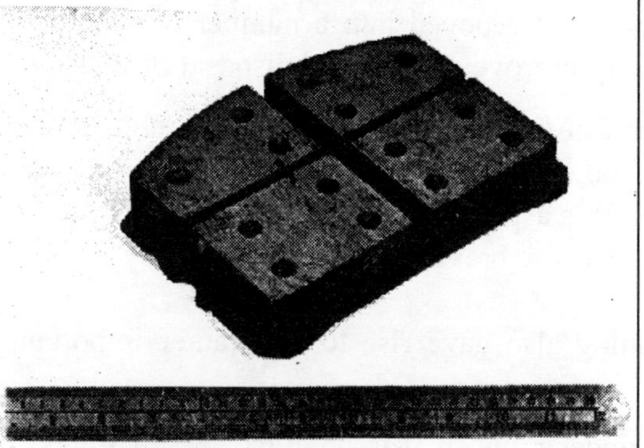

1

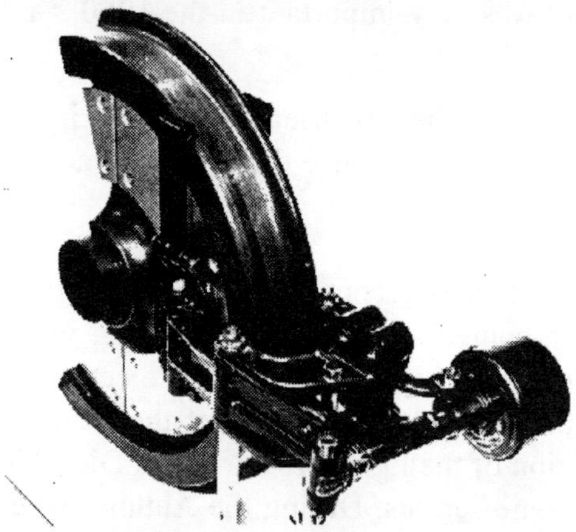

2

While railway disc brakes are [illegible] principle to those commonly used [illegible] vehicles, they are very different [illegible]

July 1976

Fig. 25 Girling railway disc brake and Ferodo disc brake pad. (Courtesy SAB Wabco BSI and Ferodo Ltd)

Brake Unit Type
BFC

Compact, light-weight design
Wide range of mounting options
Brake cylinder liner and bearings of stainless steel
Various types of parking brakes available

Fig. 26 SAB railway brake block unit. (Courtesy SAB Wabco BSI)

pneumatically operated wedge thruster fitted with linkage and brake block assembly in the form of a complete unit. These units can be used in various combinations on rolling stock, by simply bolting to the appropriate vehicle members and connecting to the pneumatic system. Apart from this particular unit, traditional block brakes still play a vital role on electric locomotives and power cars, and the fleet of 70mph diesel multiple units which are still in service on commuter and country routes. As discussed in the chapter dealing with railway brakes in Part One, the traditional material for railway brake blocks has been cast iron, which is cheap to produce and readily available. The disadvantages of cast iron blocks are :-

1. They have to be changed frequently incurring depot costs with reduced utilisation of rolling stock.

2. Cast iron dust produced during braking can give rise to serious operational and fire problems with electrical equipment, and mitigates against vehicle cleanliness.

3. Due to sparks and droplets of molten iron produced during braking, serious bogie fires can occur due to ignition of oil, sludge and accumulated debris on the bogie frame.

4. Squeal is produced during the block application especially when the train is coming to rest.

5. The surface of the wheel can sustain mechanical damage causing rumble during the brake application.

6. Lineside maintenance staff have been subjected to serious accidents due to being hit by pieces of cast iron from a disintegrating block.

Some years ago British Rail installed a brake block foundry at Horwich, originally the locomotive and carriage works of the Lancashire and Yorkshire Railway, which produced a range of cast irons with varying phosphorous content up to 3%. Brake blocks manufactured from this iron have superior wear and friction characteristics, but are more brittle and prone to cracking, necessitating block reinforcement and steel backing strips.

As discussed in the chapter dealing with friction materials in Part One, a successful organic or as it is termed composition railway block, was produced by Herbert Frood in 1907 for use on the French Metro and London Underground. A good deal of development and trial testing has been carried out over the years, and consequently this type of block has always been an alternative to cast iron for main line use. Although the initial cost is higher it has a considerably longer life than cast iron, and the friction level can be formulated to suit the particular application. This is normally higher than cast iron, which is an advantage when new stock is being designed, but a disadvantage when fitted to existing brake equipment designed for use with cast iron blocks. It has also been necessary to test a number of formulations in order to overcome problems associated with loss of rail adhesion, metal pick up in the block surface from the wheel, braking during wet conditions and thermal damage to the wheel surface. Sine the composition block is spark free it can be fitted to trains conveying petrol and other flammable substances.

The ideal application is one block per wheel with individual pneumatically operated cylinders. The ability to design the compositon block for a particular application has been recently demonstrated with the fitting of Ferodo blocks to the TGV power cars.

Railway brake actuating systems have undergone a number of changes since the introduction of pneumatic brakes in the mid 1960's. This is mainly due to increased train performance and the gradual implementation of standards originated by the International Union of Railways (UIC), and the associated Office for Research and Experiments of the International Union of Railways (ORE). This particular organisation has now become the European Railway Research Institution (ERRI).

The original Westinghouse pneumatic system as described and illustrated in Part One, *Fig.6*, enabled the brakes to be gradually applied through the drivers regulating valve, a small increase in train pipe pressure immediately releasing the brakes through the response of the triple valve. Consequent to work carried out in Europe and standardised by the UIC during the post war years, the system was modified to allow gradual release of the brakes. By also increasing the diameter of the train pipe and introducing a second pipe, these modifications enabled the system to effectively control European trains which were becoming increasingly heavier and longer.

The most fundamental change occurred with the use of the electrically controlled pneumatic valve or EP Valve, enabling electrical control of the system from the drivers cab. With the electrification of railways the use of electronic components and the application of computers, railway actuating systems have become very sophisticated. With the electrification of a major part of the UK railway system in 1966 it became possible to use electrical or dynamic braking on main line trains. Electrical braking is achieved by using the main driving motors as generators, the consequent load causing deceleration of the train. The energy created is dissipated in two ways :-

(1) By means of banks of resistors built into the locomotive referred to as rheostatic braking.

(2) By returning the energy to the supply system to be used for accelerating other trains on the same system, referred to as regenerative braking. This is a much more economical system but its use depends upon a number of factors, mainly the close proximity of other trains.

The AL6 electric locomotives introduced on the electrified West Coast route in 1966, could be used for both vacuum and pneumatic braking to suit the particular passenger stock and were also fitted with a rheostatic electrical braking system. The class 91 locomotives used on the West and East Coast routes and the Eurostar channel tunnel trains utilise both pneumatic and rheostatic braking. Both regenerative and rheostatic braking can be used on the same locomotive or electrical multiple unit, using an electronic control system to apply rheostatic braking if the line becomes unreceptive to regenerative braking. Since electrical braking is not effective at lower speeds, an electronic control system has been developed which enables electrical braking to take place at the higher speeds, with the friction braking being blended in at lower speeds for final deceleration and parking. In the event of the electrical brake failing, the friction brake must be capable of stopping the train from the maximum designated speed.

Perhaps some indication of future developments can be gained from the recent introduction of the SAB Wabco Electro Mechanical Brake Actuator, which replaces the pneumatic actuating cylinder on the disc brake caliper. The unit actuates the caliper through energy stored in a strong clock type spring, which is wound up by geared electric motors. The application and release functions are controlled electronically, the system being designed fail safe in the event of electrical failure. This is an indication of the inevitable electrical domination of the railway brake actuation system, leading eventually to the all electric train, something far beyond the vision of those early railway brake designers discussed in Part One.
Ref. (41), (42), (43), (44) (45), (46), (47) & (48)

Underground System Brakes

A brief reference to the braking systems used on the London Underground is of historical interest, since the first electric locomotives used had pneumatic brakes. These locomotives were delivered to the City and South London Railway, which was opened on 4th November 1890, from Mather and Platt, Salford Iron Works. The compressed air supply came from large reservoirs which were recharged on each trip. By 1907 there were 52 electric locomotives all fitted with air compressors. In 1922 and 1923 the Metropolitan Railway obtained 20 electric locomotives from Vickers Ltd which were fitted with air compressors for the Westinghouse brakes fitted to the locomotives, and also exhausters for the vacuum brakes fitted to the passenger stock. Pneumatic braking has remained standard practice, but the rolling stock used on the new Jubilee Line has in addition to the pneumatic brakes, electric rheostatic braking with air cooled resistors.

Herbert Frood supplied the first composition railway blocks to both the French Metro and London Underground in 1907. This type of block, unlike cast iron could be used with electrical equipment, but due to poor wet friction was confined to underground stock. In 1931 a phenolic resin bound asbestos fibre composition block was supplied by Ferodo, which could be used for both surface and underground stock.

Due to environmental reasons it has been necessary to limit the amount of sand used in the block to avoid deposits of silica dust on the station platforms.
Ref. (49) & (50)

Tram System Brakes

The traditional street tramcar was a relatively slow moving vehicle and electrical braking, block brakes and track brakes pressing on the steel rails proved adequate. With the re-introduction of the tram in the UK in the form of the Manchester Metrolink and the Sheffield Supertram, with the ability to travel at 50mph over long stretches of track, the block brakes have been replaced with disc brakes. Many cities in Europe have retained the tram system and invested in new rolling stock. This is very much the case in Strasbourg with the introduction by Adtranz of the Eurotram which is fitted with four independent braking systems comprising an electro regenerative brake on the powered bogies, electro hydraulic disc brakes fitted with Ferodo pads, spring applied parking brake and electromagnetic track brakes. *Fig.27* illustrates the disc brakes and electromagnetic track brakes fitted to the low floor bogies of this vehicle.
Ref. (51)

Fig. 27 Eurotram disc and electromagnetic brakes. (Courtesy ADtranz)

Aircraft Brakes

The history of aircraft brakes is of exceptional interest due to the severe braking requirements created both during and after the 1939-45 war, due to the vast increase in aircraft weight and speeds.

Although the internal expanding drum brake had been associated with aircraft from the earliest days, with both mechanical and hydraulic actuation it was insufficiently powerful for use with these larger aircraft. This situation was eventually overcome by using a multi disc clutch type disc brake, originally designed by Dunlop in 1935, enabling the energy to be dissipated over a number of friction faces or pairs. This concept was eventually developed for use on both military and civil aircraft as illustrated in *Fig.28*. The brake consists of a drum mounted on the wheel axle, which is internally splined to accept a number of externally splined steel discs or rotors. Sandwiched between each rotor is a steel disc or stator to which are fixed on each side a number of friction pads, the stator being internally splined and mounted on a splined stub shaft bolted to the fixed axle housing. Fixed to the outboard end of the stub shaft is a spider unit containing a number of equally spaced hydraulic pistons, fitted with automatic adjusters to maintain correct operating clearances. The brake is actuated by pressurising the hydraulic piston units, which clamp the whole unit or pack together on the splines, creating a braking force between the rotating brake discs and the static friction pad discs.

In order to provide friction material capable of withstanding the ever increasing aircraft weights and speeds, a great deal of development and testing has been carried out since the introduction of the brake. During the immediate post war years a formulation and dynamometer testing programme was carried out at Ferodo in liaison with Lockheed, in an effort to provide organic friction material for use with a multi disc clutch type aircraft brake. Even with the relatively light operating conditions of the 1950's compared to present day demands, the amount of energy dissipated during the dynamometer testing gave rise to white hot disc temperatures. Eventually organic pads were replaced with sintered metal pads, mounted on flexible stators sandwiched between flexibly designed steel rotors, the flexibility of both components ensuring minimum thermal distortion.

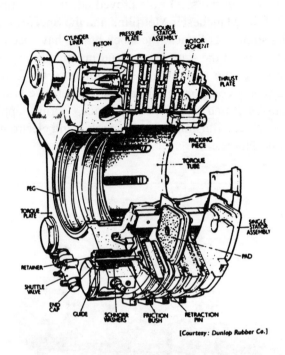

[Courtesy: Dunlop Rubber Co.]

Fig. 28 Dunlop multi-disc aircraft brake. (Courtesy Dunlop Aviation)

In order to reduce the weight and increase the performance and life of the friction material, Dunlop embarked upon the development of a carbon matrix with a carbon reinforcement known as carbon/carbon. This material was successfully demonstrated on Concorde during its introduction in 1971, and was used when Concorde entered into airline service in 1976 for both the brake rotors and stators. Since the carbon/carbon brake is 60% lighter than the equivalent size of steel brake, the weight saving was some 600Kg, the equivalent of seven passengers.

Dunlop produce both the carbon/carbon disc brake and the steel disc brake to customer requirement, the carbon/carbon version achieving 3,000 landings and the steel version 1,000 landings. In order to facilitate improved maintenance and reliability the rotor/stator assembly of the carbon/carbon brake termed the heatpack, used on the Boeing 757-200 and 757-300 aircraft can be refurbished by Dunlop on an exchange programme. This ensures a projected disc life in excess of 4,200 landings in service.

A disadvantage of the carbon/carbon material is that the density is relatively low compared to steel, and consequently the heatpack is more bulky than a metal one of similar thermal capacity. This feature has to be taken into consideration when designing the brake unit to fit into the restricted space in an aircraft wheel. Although carbon/carbon material can withstand temperatures of up to 2000°C, since the brake is adjacent to the rubber tyre and a number of alloy components, thermal insulation and cooling air passages have to be provided to prevent dangerous overheating. Dunlop design and market a brake air cooling fan to ensure low working temperatures of aircraft brakes.

The traditional Dunlop association with caliper disc brakes, referred to in the chapter dealing with the history of the vehicle disc brake is still maintained. This type of brake is now produced for braking light aircraft, helicopter wheels and rotors.

Ref. (52)

Industrial Brakes

The basic designs as described in Part One have basically remained unchanged, but there has obviously been a great deal of diversification to the present time. The application of these brakes can be briefly reviewed as follows, bearing in mind that there are many other applications :

-(1) External Contracting Band and Shoe Brakes

These are used for small applications such as spin dryer brakes and textile machines, to very large units on cranes, rolling mills, mine winding engines, hoists and winches. They arc also made in a range of sizes for use as emergency brakes.

(2) Pneumatic, Hydraulic and Magnetic Disc Brakes

These are made in a range of sizes for emergency brakes, electric motors, rolling mills and marine propeller shafts.

(3) Pneumatic, Hydraulic, Leading-Trailing Shoe Brakes

These are made in large sizes for use on tractor shovels and aircraft towing tractors.

The actuating systems are designed to suit the particular application for vehicular, plant or emergency use.
Ref. (53)

Transmission Brakes

This brake was used on road vehicles during the early 1900's usually in the form of an external shoe or band brake which clamped on a drum fitted to the transmission shaft from the engine. They were subjected to far more torque than the rear wheel brakes during deceleration with consequent deterioration of the components. When four wheel brakes were fitted the transmission brake was gradually phased out, but a modern version is still used in the form of an auxiliary brake or retarder. These are made as an electromagnetic brake or a hydraulic multi disc brake using oil immersed sintered metal friction discs. They are used to supplement the existing vehicle brake system in an integrated manner, and are usually fitted to passenger service vehicles. A transmission brake in the form of a drum brake is fited on the transmission shaft of the Land Rover.
Ref. (54) & (55)

Exhaust Brakes

The exhaust brake is used on heavy commercial vehicles to supplement the existing brake system. It takes the form of a valve fitted in the vehicle exhaust system, which can be partially closed from the cab, thereby creating a back pressure from the engine with consequent vehicle retardation.

Testing of Friction Materials

(1) Test Machines

The creation and testing of friction materials are synonymous, and when Herbert Frood began to assess various types of synthetic friction material in 1897 as an alternative to wood, leather etc. he did so using a test machine driven by a 4HP water wheel. There are no precise details of the test machine, but it can be safely assumed that it consisted of a wheel revolving at constant speed, onto the periphery of which was pressed a sample of friction material under constant load. After a fixed period of time the machine would be stopped, the sample weighed or measured for wear, and the surface condition of the sample and wheel examined for damage. In order to test the sample under wet conditions, some arrangement would have been made to apply water to the wheel surface. It is of historical interest to note that with the exception of the torque and temperature generated by the sample, the basic test criteria is the same as that currently used on brake testing machines termed dynamometers.

Frood manufactured his first commercially suitable friction material at The Herbert Frood Company, situated in premises on Green Street, Gorton, Manchester using various items of plant powered by a 20 H.P. gas engine, but there is no reference to a brake testing machine. Frood eventually moved to larger premises at Chapel-en-le-Frith, Derbyshire in 1902, and after the business became firmly established, installed in 1913 a friction material testing machine supplied by the National Physical Laboratory. This machine is illustrated in *Fig.29*, and is a more sophisticated version of the original machine made by Frood, so it would appear that it was made to his specification. These two machines were the first purpose built units made for the testing of friction materials.

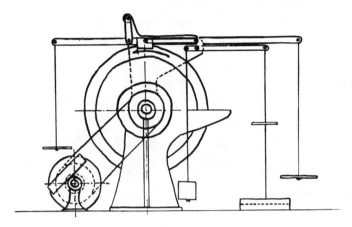

Fig. 29 National Physical Laboratory Friction Test Machine, 1913.
(Courtesy Ferodo Ltd)

Since the formation of his company Frood had, apart from commercial advertising, also produced a good deal of technical sales literature, which impressed upon the customer the importance of product testing. When the Herbert Frood Company was made into a public company in 1920 and renamed Ferodo, this same tradition continued. In order to create a new friction material that will perform safely and is commercially viable, it is necessary to carry out a formulation programme with associated testing. In the early stages a number of chemical changes will be made by the formulator, and each change will have to be thoroughly tested, and to ensure comparative results, the testing will have to be carried out on the same machine in the stable atmosphere of a laboratory or test room. This was achieved in a combined testing and chemistry laboratory which was eventually reorganised, and in 1939 a separate Test House was built as illustrated in *Fig.30*. This photograph gives some idea of the technical progress made in test machine design by the company, during its forty two years of existence. It also details some rather interesting features :-

Fig. 30 Ferodo Test House, 1940. (Courtesy Ferodo Ltd)

The machine illustrated on the right hand side of *Fig.30*, was a direct descendant of the basic test machine originally used by Frood during his early friction material experiments. It consisted of a shaft supported in bearings driven by an electric motor, to which were attached four solid metal discs. One disc was made from high carbon steel, one of medium carbon steel and two of high tensile alloy cast iron, presumably Millenite and Chromidium, as discussed in the chapter dealing with brake drum materials. Each disc was straddled by a counterbalanced lever arm pivoted on the shaft, which rested on a piston sliding in a cylinder filled with oil connected by a length of pipe to a hydraulic pressure gauge. A manually operated screw feed arrangement was fitted at the opposite end of the arm to the counterweight, which pressed a sliding shoe containing a sample of friction material onto the surface of the disc. A pipe directed water onto the disc surface for wet testing.

The test was carried out at a predetermined shaft speed and by applying sufficient load on the sample, by means of the handwheel, a moment or torque was generated over the lever arm. This pressed onto the piston creating a pressure in the cylinder which was indicated on the pressure gauge. It was part of the test machine operators job to maintain this pressure at a specified value, by occasional adjustment of the screw handwheel. The running temperature of the disc and sample at the mating surface was measured by a portable pyrometer. After running for a stipulated length of time the machine was stopped, and the samples removed for weighing and wear measurements, the surface of the sample and the disc being examined for faults and cracks.

This machine is of particular interest since it is a very early example of what is termed a Constant Torque Friction Test Machine. This concept enables a standard comparison between the samples of friction material tested, since the same output torque has been achieved independent of the friction level, by adjustment of the normal load on the sample. The sample area and the circumferential resistance created at the drum surface by the sample were calculated, such that each sample converted mechanical energy into heat at the rate of one Brake Horse Power per square inch.

This machine was used for routine control testing of Ferodo friction materials, and also for initial investigation of new experimental linings. It was not intended for duplication of actual service conditions. After being automated this particular machine was in constant use until 1956.

The test machine illustrated in *Figs.30 & 31* was designed and built at Ferodo during the early 1930's, and was of a very advanced design, possessing many features of the test machines currently in use today. It was conceived as a means of testing friction material and the associated brake assembly under actual service conditions, in terms of vehicle weight and speed. This was achieved by stopping a revolving flywheel having the same amount of kinetic energy as the moving vehicle. The size of the

Fig 31 Small brake testing machine, 1940. (Courtesy Ferodo Ltd)

flywheel used and the speed of the machine, were selected to test brakes and clutches over a range of vehicles, from motor cycles to heavy passenger cars or light lorries. The machine was driven by a 25HP motor with a top speed of 1500 r.p.m.

The machine consisted of an electric motor connected to a flywheel unit mounted on pedestal bearings. The flywheel unit was connected by a safety shear coupling to a shaft supported on pedestal bearings with a boss and faceplate keyed to the end of the shaft. The drum of the brake to be tested was mounted on the faceplate, and the brake backplate with brake shoes was fixed to a boss keyed to a shaft. The shaft end was located by a bearing in the faceplate and was supported by pedestal bearings. Keyed to the shaft between the bearings was a torque arm, counterbalanced at one end with the other end resting on a piston sliding in a cylinder filled with oil, connected by a length pipe to a hydraulic pressure gauge. The hydraulic brake being tested was connected by a flexible pipe to a hand operated hydraulic master cylinder, fitted with a pressure gauge.

In order to cover the vehicle range the flywheel unit was divided into a bank of seven smaller flywheels, one flywheel on the motor side being keyed to the shaft, the remaining six being free to revolve on shaft mounted ball bearings. Each flywheel had six equally spaced holes, through which pins of different lengths could be fitted, to connect the revolving flywheels to the fixed flywheel, the number of flywheels connected depending upon the length of the pin. An interesting feature was the counterbalance system which balances the weight of the stationary flywheels, when not connected to the fixed flywheel, thereby minimising the drag or retardation effect upon the machine.

The shear coupling consisted of a flange keyed to each shaft end, connected by a number of equally spaced copper tubes, which were designed to shear if the test brake seizes up, thereby protecting the machine from damage.

The test was carried out by selecting the number of flywheels required for the weight of the vehicle using one brake. An allowance would have to be made for rear to front weight transfer depending upon whether a rear or front brake was being tested, and also for the fixed revolving parts of the machine. The machine speed in revolutions per minute was selected to suit the speed of the vehicle. The machine was then brought up to the required speed, the motor drive switched off and the pressure applied to the test brake to decelerate the machine to rest at the rate required for a predetermined number of applications. During the period of deceleration the hydraulic pressure to the brake, and the reading on the torque arm cylinder pressure gauge would be noted, from which could be calculated the Coefficient of Friction of the friction material. The test brake was then dismantled and the linings checked for wear and condition.

The torque, brake pressure and machine speed recording instruments and the start and stop buttons were mounted together at the operators control desk, and a portable control was also provided enabling the operative to observe the test brake. A fan blower was positioned near the test brake to duplicate the vehicle cooling air, the temperature of the drum and linings being obtained by means of a portable pyrometer.

A great many variations could be carried out on the above procedure or testing schedule as requested by the customer.

Another test machine as illustrated in *Fig.32* was a much larger version of the small machine with an additional facility for testing railway blocks. It was designed and built at Ferodo in 1938, and in addition to retaining all the advance design features could also repeat the tests automatically.

The size of the flywheels and the speed of the machine were selected to test brakes for a range of vehicles, from medium weight trucks to the heaviest type of double decker bus. The total weight of the flywheels was 8 tons and the machine was driven by a 100HP electric motor with a top speed of 1000r.p.m.

The railway block unit was mainly used for assessing blocks for use on the London Underground, and was located between the two flywheel units. It was not intended to duplicate actual service conditions on the rail vehicle, being used mainly as a means of assessing the performance of experimental brake blocks. It consisted of a standard tyred railway carriage wheel keyed to a shaft, and supported by pedestal bearings, with flange couplings at each end of the shaft. Two torque frames were fitted astride the wheel being individually pivoted on bearings in the centre. The frames were fitted with hydraulic pressure cylinders and brake block holders, which were free to move radially in bearings. The torque output from each block was separately recorded, by means of hydraulic piston and cylinder units, located at the extreme ends of each frame. A water tank was positioned below the wheel for wet testing, during which a splash cover was placed over the wheel. The same testing procedure was carried out as for the small machine except on an automatic sequence.

Fig. 32 Large brake testing machine, 1940. (Courtesy Ferodo Ltd)

A great deal of intensive testing was carried out during the 1939-45 war with these test machines, in order to produce suitable friction materials for the vast amount of transport used. One of the outstanding achievements being the formulation and testing of a very successful friction material, for use on the brakes of the amphibious landing craft, which had to function efficiently under dry and wet conditions.

In the immediate post war years a great deal of investment and effort went into the design and manufacture of a small and a large test machine capable of producing constant torque results, as previously described. A reputable Swedish company designed, manufactured and commissioned the two machines, but due to lack of appropriate technology, the weight of the components and the use of extensive hydraulic systems, the necessary sensitivity could not be achieved. Consequent to these developments, it was decided to develop a constant torque control system for the standard Ferodo test machine, rather than have a dedicated machine. A good deal of progress was made in achieving this object through a liaison agreement with Raybestos Manhattan in America, who provided working details of a suitable system. It was based upon the use of bellows, which were being extensively developed at that time. The bellow is a corrugated, flexible metal cylinder, which if sealed at one end can be expanded and retracted either pneumatically or hydraulically. The unit was essentially a balancing system between the operating pressure and the torque output pressure of the test brake. It consisted of a valve system controlling the operating pressure to the test brake, which was constrained between a pneumatic reference bellows and a hydraulic bellows connected to the torque pressure generated by the test brake. If the torque pressure was above or below the resistance imposed by the reference bellows, the valve was actuated in the appropriate direction to either decrease or increase the brake operating pressure thereby ensuring a constant torque output pressure from the test brake. The unit could be suitably adjusted to cover a range of torque requirements, by altering the pneumatic pressure on the reference bellows. This system was used for constant torque testing until the eventual computerisation of the test machines.

Another form of testing which provides a basis of comparison is constant temperature testing, during which the rate of brake application is controlled to ensure that the friction material in the test brake remains at a predetermined temperature.

To assist in the formulation of friction material for use with cycles and motor cycles, a dedicated test machine was designed and installed by Ferodo in the early 1950's. In view of the relatively light torque associated with cycle and motor cycle brakes, the inherent drag effect within the machine had to be minimised. This was achieved by designing a machine with a flywheel unit in which the fly wheels not in use, could be completely removed from the shaft, and positioned for re-use as required. This was a very advanced feature which did not become common practice until the 1970's.

In 1956 a new Test House was built and the existing machines, together with a number of new test machines were installed. One of the interesting test developments that occurred at this time, was the design and manufacture of a machine to test small samples of friction material. As detailed in *Fig.33* the machine consists of a vertical spindle mounted in bearings and loaded with test weights. A flexible sample holder containing two diametrically opposed samples is held in a tapered bore at the bottom of the spindle. The samples are positioned over a cast iron rubbing ring mounted on a rotary base, which is constrained by springs and a hydraulic damper. The spindle is lowered whilst revolving at a predetermined speed, and the samples are pressed onto the cast iron surface, the resultant torque being traced on a chart recorder by a pen fitted to the base. The temperature of the samples and the cast iron rubbing ring was obtained by means of a portable pyrometer. The machine could be used on an intermittent or continuos rubbing cycle.

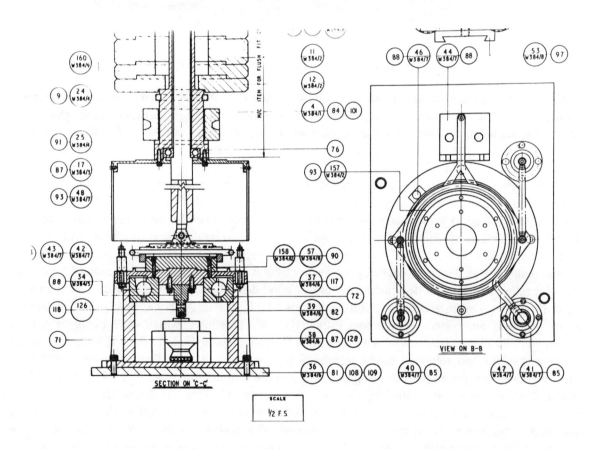

Fig. 33 Small sample test machine, 1960. (Courtesy Ferodo Ltd)

After extensive testing and statistical analysis of the results it was found that this type of test, although cheaper and easier to use than that carried out on the brake test machines, did not produce meaningful results. It was used extensively for quality control work, and is still in use in a modernised form. Quality control work is now carried out on a commercially available machine using the full size brake assembly.

Measurement and recording of the temperatures produced by the friction material has always been a very important part of brake testing. Since the late 1950's considerable development has been carried out to obtain the operating temperature of the brake lining and drum, or the disc pad and disc. This has been achieved by embedding thermocouples into the actual components, and the use of purpose designed slip rings and rubbing thermocouples, in conjunction with temperature recording equipment. A great deal of this work is now carried out using electronic sensors.

As previously discussed in the chapter dealing with railway disc brake pads, development of a suitable material started in 1954. This has developed considerably over the intervening years, with the installation of large test machines required for high speed train testing which due to their size and complexity are now supplied by specialist companies. The tests are carried out on interchangeable sub assemblies fitted to the test machine for either disc or block brake testing. These large machines are also used for the testing of heavy commercial vehicle disc or drum brakes.

It is of interest from a social history point of view that since the 1970's, apart from friction material testing, a great deal of effort and investment has been dedicated to environmental work associated with the test plant.

Herbert Frood and Ferodo pioneered a great deal of the technology used for the assessment of friction materials, based upon carefully prepared procedures in respect of both test machines and test results. The use of flywheels dedicated to the particular test, and the calibration of pressure gauges and recorders by gravity dead weights, left no margin of error. The same ideology has been realised in the automation and computerisation of the Ferodo test plant, carried out since the late 1970's. Although a great deal of the onerous routine work associated with the testing and plotting of the results has been removed, the human element still plays a vital role in the search for the best friction material.

The Ferodo Test House became the largest installation in the world dedicated to testing friction materials. It is of great historical interest that this can be traced back to the original test machine used by Herbert Frood in 1897.

Ref. (56), (57), (58), (59), (60) & (61)

(2) Test Vehicles

The ultimate test of any friction material is how well it performs on the vehicle under service conditions. This fact was very well appreciated by Herbert Frood when he purchased a 6HP Benz motor vehicle in 1899 and this vehicle would undoubtedly have been used in the development of friction material for use with band and transmission brakes. *Fig.12* in Part One shows Frood with his daughters, driving a vehicle at a gymkhana at Chapel-en-le-Frith, Derbyshire, which is situated in a hill farming and limestone quarrying region necessitating the extensive use of horse drawn heavy road vehicles. As can be seen from the advertisement illustrated in *Fig.12* in Part One, Frood was very keen to sell his patent organic brake block for use on this type of heavy vehicle. Any farmers and carters in the area who purchased and fitted the brake blocks, would undoubtedly prove to be a valuable source of data regarding their performance on the steep hills, which surround Chapel-en-le-Frith, known as The Capital of the Peak, as illustrated in *Fig.36*. The choice of such an environment by Frood proved very valuable for both sales and product development, and played a major role in the successful establishment of the business.

In pre-war days, vehicle testing would have been carried out on cars, lorries, buses and trains in liaison with the customer, apart from Ferodo staff cars and delivery vehicles. During the immediate post war year, dedicated vehicle testing was introduced, and a fleet of cars, lorries and buses was gradually acquired, and eventually housed in a Test Garage. Road testing was carried out to Ferodo Test Schedules in the hill country of Derbyshire and the relatively flat Cheshire Plane, in order to test the materials over a wide range of operating conditions. Dedicated testing for customer submission was also carried out, and with the growth of the private car eventually became the main testing requirement. Specialised tests which could not be carried out on the public roads were done at the Motor Industries Research Association proving ground at Nuneaton. This was established on a wartime airfield, and during these early days consisted of a number of runways with a small collection of huts. High speed braking stops were carried out on the runways, the deceleration being measured by a portable Ferodo/Tapley pendulum decelerometer. The braking distance was measured by marking the runway with white paint, manually applied through a hole in the test vehicle floor, at the start and finish of the brake application. This procedure was eventually improved by replacing the paint brush with an air pistol, which fired a piece of white chalk on to the runway. It is of historical interest to note the rather primitive testing methods used in the immediate post war years due to the lack of suitable technology.

With the growth of the electronics industry, it became possible to design and install dedicated instrumentation for measuring and recording the deceleration, the brake pipe line pressure and the operating temperature of the friction material and the drum or disc. *Fig.34* illustrates the instrumentation fitted into the test vehicle interior in the early 1980's. *Fig.35* illustrates a test bus on the Cheshire Circuit, loaded with concrete ballast blocks, to simulate the passenger weight. These concrete blocks were also fitted into the test vehicles but were eventually replaced by test dummies, filled with the appropriate weight of water.

As was the case with the test machines, vehicle testing has been computerised, with the computerised recording unit located in the car boot or lorry cab, in the form of an aircraft black box. Sine the results can be plotted and analysed at the completion of the testing programme, this equipment has been invaluable for tests carried out on the Grössglockner, and other European testing locations. Due to the intense competition now prevalent in the car industry, this kind of testing is very important. It is carried out in liaison with both the brake and vehicle manufacturers and a number of friction material suppliers competing for the same business. At the completion of the tests, the brake shoes or disc brake pads are removed from the vehicle for measurement and examination, the data being filed with the relevant computer records.

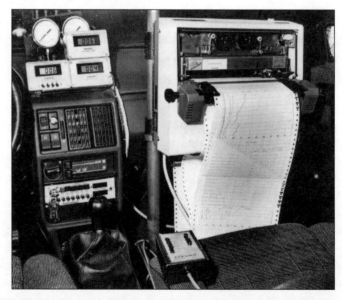

Fig. 34 Brake test vehicle instrumentation, 1980. (Courtesy Ferodo Ltd)

As discussed previously, the friction material manufacturer has always maintained strong links with car and motor cycle racing. This dates back to pre war years both in the UK and Europe with Ferodo linings being used at Brooklands in 1911, and the T.T. Races in 1912, the start of an association with the racing world which is still maintained to the present day. It was done not only to gain publicity for the product, but to also test performance under very demanding conditions, with a dedicated staff and a mobile workshop being used in the 1950's and 1960's. A good deal of valuable data has been obtained which is used in the formulation of friction material for public use, hence the term racebred material.

Ref. (62), (63), (64) & (65)

Fig. 35 Brake test passenger service vehicle, 1960s. (Courtesy Ferodo Ltd)

The Friction Material Industry

The friction material industry both in the UK and throughout the world has always been, and still is specialised and secretive. Due to the nature of the product in respect of formulation and manufacture this is understandable, and a good deal of testing and analysis of competitors materials has always been a feature of the industry. This is the reason why the bulk of the information contained in this section is based upon the verified industrial records of Ferodo Ltd., the pioneers of the product in 1897. The invention of organic friction material by Herbert Frood, its early application and development have been recorded in Part One. During those early pioneering days a relatively cheap and simple medium was devised, with the ability to convert the kinetic energy of a moving vehicle into heat, whilst being brought to rest. Friction material is the vital interface between the static and rotary parts of the braking system, and its failure leads to catastrophic results for both passenger and vehicle. This aspect has been foremost in the mind of the manufacturer, and consequently the UK friction material industry has an excellent record, although inferior and dangerous materials have been imported from other parts of the world.

With the increase in private and commercial transport prior to the 1914-18 war, Frood responded by providing friction material to satisfy the higher speeds and weights. This was achieved by the formulation of higher temperature impregnants, and eventually in 1908, the replacement of cotton fibre by asbestos. During the war years a completely new material was introduced, capable of withstanding the arduous conditions imposed when steering and braking a 40 ton tank on the battlefield. This particular application necessitated the development of a special resin or binding substance, capable of holding the asbestos and other components together under conditions of extreme heat and oil contamination. By 1924 it had been developed into a friction material suitable for commercial application. In order to provide brake blocks for motor cycles, Frood introduced the first moulded friction material. It was produced by what is termed a vulcanising process, the various constituents being mixed with a rubber binder and subjected to heat and pressure in a die.

It is of interest at this particular time to put into context the previous reference to specialisation. It must be borne in mind that Frood had pioneered a new industry requiring new technology in terms of chemistry, manufacturing and testing. An examination of a group photograph taken at the Herbert Frood Company in 1918, shows that there was a total workforce of 130 people covering all aspects of the business, with the exception of plant manufacture which had to be contracted out. It was fortunate for Frood that there existed in Chapel-en-le-Frith a cast iron foundry, and also an engineering works owned by a Mr John Chaloner, which had been founded in 1770, and had provided an engineering service to a number of local cotton mills. During these early formative years Frood and Chaloner conceived and manufactured the first specialist plant for the manufacture of friction materials, sub-contracting out the casting work to the foundry.

This business relationship lasted many years and is a classic example of the growth of a service industry. A similar situation also came about at this time with the establishing of another friction material factory at Buxton, Derbyshire, owned by a separate company eventually trading under the name of Duron. There was still a connection with Frood through the marriage of one of his daughters into the family who owned the business. The company built up a very good commercial vehicle lining business with the local lime haulage industry, and is now part of the T & N plc Ferodo Friction Group.

In 1920 the Herbert Frood Company became a public company and was renamed Ferodo, being an anagram of Frood's name and that of his wife Elizabeth, a name that has become synonymous with friction material. In 1926 Ferodo became part of Turner & Newall, who had extensive asbestos mining operations in Canada and Rhodesia. Frood retired from Turner & Newall in 1927, and settled in Colwyn Bay where he died in 1931, at the age of 67.

During the early 1920's friction material was being supplied to all the well known vehicle manufacturers such as Austin, Crossley, Hotchkiss, Lagonda, Ford, Leyland and Morris, and with the introduction of four wheel braking, the demand increased. In 1922 the Austin 7 was the first mass produced car to be fitted with four wheel brakes, using Ferodo brake linings as original equipment. With the growth of the brake manufacturing industry, friction material was also being supplied for use with the Girling mechanical and Lockheed hydraulic leading-trailing shoe brake, and the Bendix duo-servo shoe brake. Material was also available in roll form for cutting and riveting to brake shoes with Ferodo clamps, anvil and punches.

During the early 1900's Frood had carried out a good deal of development work for the provision of linings for colliery winding engines and cranes, replacing the wooden or metal blocks with organic material. Since 1907 composition railway blocks had been supplied to the London Underground and the French Metro.

With the increasing popularity of motor sport, friction material was supplied for all the competing cars, and the world speed record of 169.23mph achieved by Parry Thomas on Pendine Sands in BABS, used Ferodo friction material. Frood had always advertised his products well, and apart from racing produced technical literature and general information of interest to the motorist. As a result of this policy Ferodo became a household name, with advertisements in the popular press, motoring magazines and on selected railway bridges. Some years later in 1976 it was listed in the Worlds Top 12 of classic invented names. As part of a continuing association with the motoring public, Ferodo published a brake performance chart showing the relationship between the stopping distance of a car travelling at a given speed, and the percentage efficiency of its brakes. This chart also tabled suitable retardations for a given vehicle, and was used in conjunction with a Ferodo pendulum type decelerometer made by Tapley. This is an interesting historical development since on 31 May 1937, the Ministry of Transport introduced new braking regulations. It was made a legal obligation for motorists to allow any uniformed police constable to test and inspect their brakes. In view of the new Ministry of Transport regulations Ferodo took the initiative and opened Brake Testing Service Stations in liaison with a number of garages. An advertising campaign was launched and motorists participating in the scheme were issued with a dated certificate verifying the efficiency of their vehicle brakes. By 1950 there were over 4000 Ferodo Brake Service Stations, predating the existing MOT certificate by many years.

During the 1939-45 war a vast amount of friction material was produced for the war effort, including a special formulation based upon railway composition blocks, for use in wet and dry conditions on amphibious vehicles.

In order to satisfy the post war boom, as previously described, in private and commercial transport. both research and manufacturing facilities were extended. It was at this time that a very serious problem was experienced by a considerable number of motorists, due to partial or complete loss of braking at higher speeds or during periods of heavy brake usage. The solving of this problem, which became known as brake fade, was obviously treated with great urgency. It was eventually found to be caused by high temperature melting of the surface of the brake lining, due to the increase of vehicle performance and speed. The melting caused loss of grip and was overcome by developing a binder resin with a much higher melting point. In order to differentiate between linings made from this formulation and the previous material, the new product was commercially launched as Ferodo Anti Fade brake linings.

In the 1950's a rubber based formulation was developed to produce a flexible roll of friction material. This was supplied to garages for cutting and fitting to the various brake shoes without having to be radiused.

During the 1960's, in common with the rest of industry where applicable, group technology was introduced with the changing of many long established working practices. In order to economically increase production the mixing of the various constituents was carried out in a large central mixing plant, rather than a number of smaller ones. Formulation changes were introduced enabling the material to be rolled and a continuous sheet could be :-

1. Cut to a predetermined length.

2. Radiused in rollers to the required diameter.

3. Baked in high temperature ovens.

4. Ground on the inside and outside radius.

5. Cut to the required width.

The introduction of disc pads for vehicles and railways during the 1960's necessitated a good deal of plant design and formulation testing.

The sequence of manufacture was :-

1. Manufacture of a pad preform in a compacting die.

2. Baking in high temperature ovens.

3. Finish grinding and grooving as required.

Disc pad manufacture is now a very complicated process and is carried out automatically in computer controlled robotic cells.

There have been two major events in the history of the friction material industry, the first was the introduction of asbestos in 1908 and its removal in the 1980's. The word asbestos comes from an ancient Greek word meaning unquenchable due to its heat resistant properties. It was mined in Italy and Corsica in early times, and was obviously a very much sought after mineral for high temperature applications, and was an ideal choice for use in friction material. It was easily available and therefore cheap and it became an irreplaceable component over a period of seventy years. In the post war years there was an awareness of the associated health risks during the manufacturing stages. The responsible manufacturers within the industry installed a great deal of extraction equipment, modified working practices, and introduced routine dust and medical checks. In spite of these precautions, and due to the negligence of certain companies, the environmental danger was considered to be too great. In order to prepare for this fundamental change to the product, a research and development programme was carried out at Ferodo in the late 1970's with a view to creating asbestos free friction materials. This was achieved in 1980 with the first European contract to supply asbestos free disc brake pads for the Morris Ital. Other non asbestos contracts followed for Ford, Jaguar and Austin with all material being asbestos free by 1989. The introduction of asbestos free friction materials was far from trouble free both in terms of manufacture and performance, and its introduction during the 1980's coincided with the serious problem of disc thickness variation (DTV). This phenomenon became apparent as a result of the exceptional number of warranty claims filed against the vehicle manufacturers at this time for vibration in the steering column. After a good deal of investigation by both vehicle and brake manufacturers the trouble was diagnosed as a minute variation in the brake disc thickness. During the application of the pads for full or partial braking, high frequency pulses caused by the disc thickness variation were transmitted through the suspension to the steering mechanism and car interior. Two basic explanations were considered :-

1. The relatively compatible absorbent asbestos pad had been replaced by a non resilient material, containing a good deal of metal and nylon fibre that was incapable of absorbing the pulses.

2. Due to the extension of the motorway system, the pattern of braking had changed, since brakes were only applied for the occasional high speed partial application, the pads rubbing continuously against the disc for most of the journey.

Since the variation in the thickness of the brake disc was very small, it was necessary to fit very fine electronic probe measuring equipment to the test dynamometers and vehicles. The material formulators were then able to ascertain the effect of new formulations upon the disc surface. Fine measuring equipment was also designed to check each test disc, which was individually identified, and a precision brake disc turning and grinding facility was obtained. More amenable formulations have been produced but the problem is somewhat ongoing and the use of this equipment is now part of the routine test procedure.

In view of the total comfort conditions imposed by the car manufacturer, it is very important that modern friction materials provide a smooth braking action. This is now necessitating the use of special purpose noise detection dynamometers, capable of accepting the test brake assembly complete with suspension unit.

For reasons of economy it is essential that the component supplier is sited as near to the vehicle manufacturer as possible, and consequently Ferodo factories as a part of the T & N plc Ferodo Friction Group have been established on a world wide basis, a trend which is set to continue with the setting up of vehicle manufacturing plants in South America, India and China.

In 1897 Herbert Frood created the first factory in the world to create quality organic friction material, to suit the requirements of a wide range of customers. *Fig.36* is a photograph of the Ferodo factory in the 1960's with a fine view of the Derbyshire hills, making an interesting comparison with Frood's first premises as illustrated in *Fig.10* of Part One. He would indeed be grateful to know that one hundred years later, this same tradition continues with a current annual production of some 30 million car disc pads, 12 million linings and a considerable quantity of commercial vehicles disc pads, railway disc pads and blocks for customers world-wide.
Ref. (66), (67) & (68)

Fig 36. Ferodo Ltd, Chapel en le Frith in the mid 1960s with the Derbyshire hills in the background. (Courtesy Ferodo Ltd)

The Past and the Future

Many of the developments discussed in the Conclusion section of Part One have materialised. Constant advances have been made in the technology of brakes and friction materials in response to the ever increasing performance and speed of the vehicle. This has been the case since earliest times and is set to continue in many and various ways. The use of lighter materials for the brake components, including ceramic coated or carbon fibre discs for road vehicles and trains. The increasing use of computers and electronic components in the various brake systems, leading to braking by wire, initially for trains and commercial vehicles and eventually for cars. The changes brought about by all electric vehicles will be far reaching with the eventual development of low speed electric braking.

In the light of this particular paper it is a source of comfort, to safely assume that the friction brake will still be required, even if only for the lower speeds, emergency use and parking. Some idea of the future development of friction materials is perhaps given by the use of carbon fibre pads on the disc brakes of the recently introduced Thrust SSC turbo vehicle. On the 13th October 1997 this vehicle became the first in the world to attain a supersonic speed of 764.168mph and on 15th October 1997 a new world land speed record with an average speed for the two runs of 762mph.

Some years ago the Institution of Mechanical Engineers produced a slogan which stated that nothing moves without mechanical engineers, perhaps it is also equally true to state that nothing stops without mechanical engineers!

Particular & General References

The Ferodo International Technical News data sheets have been referred to as F.I.T.N. with the appropriate reference number and date of publication. The Institution of Mechanical Engineers publications are available from the Institution Library, and the source of the remaining references is indicated as appropriate.

Internal Expanding Drum Brake

1. Leading-Trailing Shoe Brake
(1)   "Automobile Brakes and Braking Systems" by T.P.Newcomb and R.T.Spurr.
        Motor Manuals No 8, Chapman & Hall, 1969.
        Ref : Chapter III and Appendix III.
(2)   F.I.T.N. A6 - "Girling Drum Brakes for Cars", November 1971.

2. Duo-Servo Shoe Brake
(3)   "Automobile Brakes and Braking Systems" by T.P.Newcomb and R.T.Spurr.
        Motor Manuals No 8, Chapman & Hall, 1969.
        Ref : Chapter III and Appendix III.
(4)   F.I.T.N. A15 - "Perrot Drum Brakes", November 1974.

3. Two-Leading Shoe Brake
(5)   F.I.T.N. A6 - "Girling Drum Brakes for Cars", November 1971.
(6)   F.I.T.N. A14 - "Lockheed Car Drum Brakes", July 1976.

4. Relative Performance of Drum Brakes Described
(7)   "Automobile Brakes and Braking Systems" by T.P.Newcomb and R.T.Spurr.
        Motor Manuals No 8, Chapman & Hall, 1969.
        Ref : pp72-74.

Also of interest :
- "Vehicle Braking" by A.K.Baker, Pentech Press, 1986.
- "Automobile Brakes and Brake Testing" by M.Platt, Pitmans Automobile Maintenance -Series, 1968.
- "Commercial Vehicles, Engineering and Operation" I.Mech.E.1967.
- Ref : pp153-176.
- Girling & Lockheed brake servicing manuals.

Disc Brakes

1. Automotive Disc Brakes
(8)   "The Ferodo Story 1897-1957" Ferodo Ltd.
(9)   The Motor, 1st October & 8th October 1952. Haymarket Publishing.
(10)   F.I.T.N. A1 - "Swinging Caliper and Others", January 1969.
(11)   F.I.T.N. A2 - "Disc Brake Pads", July 1969.
(12)   F.I.T.N. A3 - "Discs versus Drums", March 1970.
(13)   F.I.T.N. A5 - "Lockheed Disc Brakes", July 1971.
(14)   F.I.T.N. A10 - "Girling Disc Brakes", March 1973.
(15)   F.I.T.N. R12 - "New Light on Cast Iron", March 1976.
(16)   F.I.T.N. A23 - "New Girling Calipers", March 1977.
(17)   F.I.T.N. A31 - "Disc Brake Evolution", November 1979.
(18)   F.I.T.N. R24 - "Cast Iron and Titanium", July 1980.
(19)   "Cast Iron Brake Rotor Metallurgy" by B.J.Chapman and D.Hatch.
        I.Mech.E - I.R.T.E. Joint Conference, University of Loughborough, March 1976.
        Ref: pp143-152.

Also of interest :
"Automobile Brakes and Braking Systems" by T.P.Newcomb and R.T.Spurr.,
Motor Manuals No 8, Chapman & Hall, 1969.
Ref : Chapter IV.

2. Commercial Vehicle Disc Brakes

(20)   "Vehicle Braking" by A.K.Baker, Pentech Press, 1986.

3. Motor Cycle Disc Brakes

(21)   F.I.T.N. A33 - "Some Motor Cycle Disc Brakes", July 1980.

4. Tractor Disc Brakes

(22)   F.I.T.N. A19 - "Girling Tractor Brakes", November 1975.

Vehicle Actuating Systems

1. Mechanical

Private Car

(23)   "Automobile Brakes and Braking Systems" by T.P.Newcomb and R.T.Spurr.
         Motor Manuals No 8, Chapman & Hall, 1969.
         Ref : pp208-228.

2. Hydraulic and Pneumatic

Private Car

(24)   F.I.T.N. B2 - "Hydraulic Brakes", November 1969.
(25)   F.I.T.N. B5 - "Easier Stopping", November 1971.
(26)   F.I.T.N. B10 - "Separate Circuits for Safety", March 1974.
(27)   F.I.T.N. B12 - "Warning the Driver", November 1974.
(28)   F.I.T.N. B27 - "Girling Master Cylinders", March 1979.

Also of interest :
- "Automobile Brakes and Braking Systems" by T.P.Newcomb and R.T.Spurr.,
- Motor Manuals No 8, Chapman & Hall, 1969.
- Ref : Chapter IV.
- "Vehicle Braking" by A.K.Baker, Pentech Press, 1986.
- "Automobile Brakes and Brake Testing" by M.Platt, Pitmans Automobile Maintenance Series, 1968.
F.I.T.N. B23 - "Citroen Hydraulic Brake Systems", November 1977.

Commercial Vehicles
(29)   F.I.T.N. B8 - "Air Brake Systems", July 1973.
(30)   F.I.T.N. B11 - "Air Brake Systems 2", July 1974.
(31)   F.I.T.N. B13 - "Air Brake Systems 3", March 1975.
(32)   F.I.T.N. B14 - "Girling Skidcheck", July 1975.
(33)   F.I.T.N. B16 - "Maxaret for Trucks", March 1976.
(34)   F.I.T.N. A32 - "Kirkstall Cam Brakes", March 1980.
(35)   F.I.T.N. B24 - "Braking Medium Weight Trucks", March 1978.
(36)   F.I.T.N. B28 - "Bendix Anti-Lock", July 1979.
(37)   "Commercial Vehicles, Engineering and Operation" I.Mech.E.1967.
         Ref : pp153-176.

Also of interest :
- "Vehicle Braking" by A.K.Baker, Pentech Press, 1986.
- "Braking of Road Vehicles" I.Mech.E. March 1993.
- F.I.T.N. B15 - "Air Brake Systems for Trailers", November 1975.
- F.I.T.N. A8 - "Lockheed Truck Brakes", July 1972.
- F.I.T.N. A17 - "Girling Truck Brakes", March 1975.

Motor Cycle Brakes
(38) F.I.T.N. A33 - "Some Motor Cycle Disc Brakes", July 1980.

Brake Fluid
(39) F.I.T.N. B4 - "Hydraulic Brake Fluid", March 1971.
(40) F.I.T.N. B31 - "Brake Fluid Development", July 1980.

Railway Brakes
 (41) "Railway Braking, A History" by A.C.Sharpe, Engineering. September 1979.
(42) Report on the Collision, Chapel-en-le-Frith (South), H.M.S.O. 1957.
(43) "Development of the Disc Brake with Particular Reference to British Railway Application" by J.B.Tompkin. Journal of the Institution of Locomotive Engineers, February 1969.
(44) "The High Speed Train" by B.G.Sephton, I.Mech.E. Railway Division Journal, September 1974.
(45) "Brakes along the Line" by M.J.Leigh, I.Mech.E. Railway Division, September 1992.
(46) "Euston Main Line Electrification". I.Mech.E., October 1966.
(47) "Composition Brake Blocks and Tyres" by S.Wise and G.R.Lewis, I.Mech.E. Railway Division, 1970.
(48) "Composition Friction Materials for the Replacement of Cast Iron Railway Blocks" by P.A.Gibson, Railtech I.Mech.E. 1996.

Underground Brakes
 (49) "The Story of London's Underground" by J.R.Day, London Transport Pub's, 1974.
(50) "Composition Friction Materials for the Replacement of Cast Iron Railway Blocks" by P.A.Gibson, Railtech I.Mech.E. 1996.

Tram Brakes
 (51) Technical data from the National Tramway Museum, Crich, Matlock, Derbyshire.

Also of interest :
- "Eurotram - 100% Low Floor Tramcars for Strasbourg" by K.Brereton, I.Mech.E. Railway Division (Midlands Centre) Light Rail Colloquium 22nd March 1995.

Aircraft Brakes
(52) Technical data from Dunlop Ltd. Aviation Division.

Also of interest :
- "Mechanical Engineering, A Decade of Progress 1960-1970", I.Mech.E. 1971. (Ref : p171)

Industrial Brakes
(53) F.I.T.N. Ref : 12-159.

Transmission Brakes
 (54) "Automobile Brakes and Braking Systems" by T.P.Newcomb and R.T.Spurr., Motor Manuals No 8, Chapman & Hall, 1969. Ref : pp202-213
(55) F.I.T.N. A27 - "Ferodo Retarders", November 1978.
(56) "Development of Electromagnetic Retarder Control Systems" by R.F.Gregg. Ref : C444/052/93 Braking of Road Vehicles, I.Mech.E. 1993.

Testing of Friction Materials

1. Test Machines
 (57) F.I.T.N. R1 - "Ferodo Test House", January 1969.
(58) F.I.T.N. R5 - "Inertia Test Machines", November 1970.
(59) F.I.T.N. R10 - "Small Sample Testing", March 1974.
(60) F.I.T.N. R20 - "Testing the Product", March 1979.
(61) F.I.T.N. R25 - "New Inertia Test Machine", November 1980.

<u>2. Test Vehicles</u>

(62)   F.I.T.N. L2 - "Measuring Deceleration", July 1971.
(63)   F.I.T.N. R4 - "Vehicle Brake Testing", March 1970.
(64)   F.I.T.N. R6 - "Vehicle Test Schedules 1", March 1971.
(65)   F.I.T.N. R7 - "Vehicle Test Schedules 2", November 1971.

The Friction Material Industry

(66)   F.I.T.N. F13 - "Friction Materials", March 1970.
(67)   F.I.T.N. R9 - "Quality Control", November 1972.
(68)   F.I.T.N. R23 - "Ferodo Tables of Stopping Time and Distance", March 1980.

Also of interest :
- "The Ferodo Story 1897-1957" Ferodo Ltd.
- The Derbyshire Countryside, August-September 1956 pp35-39.
- "Friction Materials : Black Art or Science?" by H.Smales, I.Mech.E. 1994.